IMPROVE**IT!**

A collection of essays on using analytics to accomplish more with SharePoint

Loren Johnson, editor
Product Marketing, Webtrends

Jean-Marc Krikorian, editor
Partner Alliances, Webtrends

Mark Miller, publisher
End User SharePoint

ES˙
EndUserSharePoint.com

"What a wonderful collection of experiences . . . This was a joy to read! Not only are these authors providing great experiences, best practices and lessons learned, but these are recognized industry experts like Naomi Moneypenny, Christian Buckley, Jeff Shuey, and Richard Harbridge. There's real impact in the things they've shared. There's great guidance and understanding in impact of measuring to deliver real value. One of the most important things in any deployment is understanding adoption and usage then applying changes for improvement. The line up of authors is truly incredible. Keep this one close, I'm sure you'll love it!"

Joel Oleson
Enterprise Community Leader & Information Architect

IMPROVE IT!
A collection of essays on using analytics to accomplish more with SharePoint

Webtrends
851 SW 6th Ave
Portland, OR 97204
www.webtrends.com

Interior design by Hillspring Books, Inc.

ISBN 978-0-692-38357-5

Printed in the United States of America

Contents

Foreword

"I have no idea what you just said.
What does that mean?"

You'd be surprised how often in a day I hear this sentence, or one just like it. I love the candor it displays. Truthfully, I probably just need to keep working on my communication skills! In technology, we throw around terms—and often simply their acronyms—and expect others to know what we mean. This happens so often that we have a glossary of terms on the home page of our intranet that I personally feel grateful for.

For years we've heard about data and its uses, without deeply understanding what that means. From content consumption figures, to user journeys through your customer experience, to metrics from your automated process, we have the ability to gauge exactly what is happening across these segments of our business. But data alone is not enough.

We need meaning and insights from that data in timely ways. Attaining significant meaning—in communication, data analysis or even with our loved ones—is not a fly-by-night operation. It takes intention, discipline, investment and patience. It is a journey in which you make a commitment to being teachable and open-minded about where the data may take you. Deriving meaning is a multi-dimensional, non-linear activity.

The collaboration and productivity industries have long been plagued by a lack of meaningful data analysis within organizations. Today, business is impacted by a need to derive and communicate what it gains by investing in metrics. CxOs are rightfully asking, "What are we getting from this?" It is no longer enough to measure key indicators — and truthfully it never was.

The authors of this book can guide you in turning this never-ending stream of data into business insights. These insights can help you make real-world decisions that will impact the success of your initiative. But we can achieve this only when and if we are willing to invest not just in data capture and analysis, but in the journey to attain meaning.

It is important to note that data is not a substitute for experience or intuition. Human insight based on complex-pattern recognition of the ecosystem around us is a survival skill hard-wired into our brains. But like all innate capabilities, it can fall into ruts of perception about situations, technology, people and ourselves.

You may be confronted with the opportunity to completely rethink your corporate intranet. The question is not how but why, and what outcome you desire. Your experience and intuition can be validated by analyzing your current state with the right data at the right time.

Getting the right data at the right time is no small feat. What is exciting about technology today is the ability to generate near real-time and streaming data, across experiences and devices. This can be analyzed by any number of criteria to prove or disprove your "hunch" about what you should do next. Business insights are not derived from the data itself, but from our interaction with it. Interpretation based on our own business acumen is critical to attaining meaning.

In the pages of this book you will find a variety of ways to attain meaning and leverage data to derive business insights from your collaboration experience and beyond. I'm honored to be among the people participating in this project. I consider these authors, many of whom I know personally, to be thought leaders in their respective areas. Collectively they represent best practices, innovative insights and decades of experience.

What I encourage you to do is take charge of your data. Take that first step toward the multi-dimensional learning that comes from your direct interaction with this source of insight. Release your preconceived notions about what it will teach you, so you become teachable and knowledgeable at the same time.

Embrace the challenge coming from your C-suite. Let it inspire you to:

- Prioritize the definition and measurement of key indicators.
- Champion the investment in foundational data tools.
- Drive innovation in creating meaningful collaboration experiences.
- Share your knowledge with other people trudging the path of happy destiny toward meaning.

As our collaboration experiences expand beyond each other toward the Internet of Things, this capacity to derive meaningful insight from data will be in even greater demand. But first, let us have a deeper understanding of what is valuable about the experience we are having today.

I hope you enjoy this book and share your experiences with us online. We stand ready, willing and able to help you along this journey!

Karuana Gatimu
Director, Engagement & Adoption
Founder, SharePoint Community Leadership Board
Microsoft IT
@Karuana

Introduction

Measuring Employee Engagement as It Matures over Time

by Natalie Hardwicke

Employee engagement is an interdisciplinary and often misunderstood practice. Its relevance and measurement has been debated among business professionals, academics and psychologists for more than 25 years. The measurement of employee engagement has exploded in the age of the digital workplace, when employees can be connected, informed and monitored anytime, anywhere, and across multiple media devices.

Despite the lens of analysis that is possible thanks to digital measurement of employee behavior, the biggest predictors and drivers of employee engagement have remained stable over time. These include the following five disciplines:

- Organizational culture
- Leadership
- Communication
- Knowledge management
- Emotional intelligence of employees

This foreword provides insight into these five disciplines, and discusses how desired behaviors can be observed and measured as an organization matures over time.

Why Engagement?

Engagement is when employee drive meets organizational need. It's when an individual employee has a personal and emotional investment in their work, their leaders and the vision of the organization. Having employees at a stage where they are considered "engaged" is a reflection of organizational growth and an employee commitment to that growth.

In my own professional career I have had to convince business leaders of the importance of employee engagement. And in all cases, I have been asked to provide evidence as to why the business should invest more time, attention and money in their employees.

In 2013 I was working as a communication and employee engagement manager. One of my main roles was the management of a SharePoint platform for 3,000 employees. On the platform, my team published a variety of messages, documents and information. Some messages were need-to-know messages that directly impacted the work of employees, and some messages were the sharing of good news and success stories relating to employee achievement and completed work. When I looked at the analytics for who was looking at what content, I made three main observations. Firstly, the need-to-know messages were only being read by team leaders. Secondly, employees in three geographical locations were not visiting the platform. And lastly, employees were clicking into blogs and discussions, but no one was commenting or providing feedback.

These are the findings that the analytics told me, and it was through the analytics that my investigation began. In observing the team leaders, I discovered that the need-to-know messages were being cascaded to junior employees in team meetings. Even though employees weren't reading the messages themselves, they were still receiving the message content. These employees felt that messages of importance should come from their direct leader. In relation to the three non-participating locations, I discovered that employees in these sites were either on the phone to clients all day, or they were responsible for responding to client requests, which involved the use of three different software products. This meant their time and attention was limited. To investigate the employees not commenting or contributing to the platform, I conducted some focus groups. I found that employees didn't know if they were allowed to comment, didn't really understand how to use the platform, and had never before had to provide feedback via a digital and social channel.

This process taught me that analytics can be used as a basis for investigating employee behavior, beliefs and practices. But most importantly, it taught me that these measurements were specific to the time of year and the culture of the business. Engagement is something that matures and changes as the business itself matures and changes.

In terms of the five disciplines (culture, leadership, communication, knowledge management, and emotional intelligence), culture reflects community or "the way we do things around here." Communication cannot be successful without good knowledge management, because communication is often the product of contextualized information that stems from a body of knowledge. The emotional state of employees can be directly influenced by the actions and competencies of their leaders, as they are the people who decide what tasks the employees work on, and to some extent, how the employees work on these tasks. Leaders decide on and approve the organization's vision. This includes what systems and software employees can use in their job, what projects the organization will focus on, and what investment the organization is going to make in communicating with and developing its people — all of which becomes the bread and butter of workplace culture.

Who Owns Engagement?

Employee engagement has been found to reduce rates of employee turnover, to enable better customer service with clients and external audiences, and even to cause employees to turn down job promotions just to stay working in a particular team or for a particular leader. However, the measurement of employee engagement as it matures over time can be inherently complicated, especially because communication, leadership and emotional intelligence are somewhat subjective.

Traditionally, employee engagement has been measured on an annual basis through the use of a survey, which is normally managed by the Human Resources (HR) department. The traditional engagement survey is not effective as a singular form of measurement, because it doesn't assess the pulse of the organization over periods of change or fluctuations that happen throughout a calendar or financial year.

In the modern workplace, the ownership of measurement has expanded to include communication and IT teams, leaders, individual employee responsibility, and the incorporation of external client feedback. The measurement can be constant, thanks to analytics. And the modern application of the multi-method approach to measurement means that organizations can use the findings to invest in improvements. Engagement is owned by everyone, and it matures at different speeds and competencies. Analytics can be measured daily, but their usage in forming employee questions should be reserved to specific times of year (such as every quarter) or during known peaks and troughs. Whenever engagement is measured, action needs to be taken based on those results. You can then look at how those results have been implemented the next time you measure engagement.

Measuring the Five Maturity Stages

The employee engagement maturity model reflects the five influential disciplines (culture, leadership, communication, knowledge management, and emotional intelligence) and shows what behaviors are evident from these disciplines as the organization matures over time.

The five stages of maturity defined in this model are:

1. **Non-existent:** In this stage, the organization has identified a need for change.
2. **Initiated:** The organization invests in research and exploration. Processes begin to emerge.
3. **Defined:** The organization defines a baseline and strategy for its engagement plan. Collaboration is actively pursued.
4. **Aligned:** This is the adoption stage, in which systems, work areas and people are integrated.
5. **Enacted:** The organization operates as a single cohesive unit, but remains adaptive to change and continual improvement.

Using the employee engagement maturity model as a framework, let's explore the stages in detail and discuss measurement techniques for each. This measurement focuses on behavior, processes, perception and analytics. Along with the information, you'll find advice on how you can measure stages of maturity in your organization and what you can do with your findings in order to progress along the maturity scale.

Fact-Check in Stage 1

Some aspects of measurement are based in fact. For example, if your enabling business functions are not part of the same unit, you don't need to measure that, you need to change that. If your organization has no avenue to capture employee voice via a channel that isn't associated purely with the employee's direct leader, then you need to introduce a new channel. You don't need to measure the "what ifs" of these scenarios; numerous academic papers and organizational case studies have already done that for you.

One of the fundamental aspects of measuring engagement is ensuring you have platforms or situations that provide an avenue for measurement. Stage one involves setting up processes that allow you to measure engagement as it matures. Stage one may represent a new organization, or it may be evident in new employees or even certain teams who are disengaged. This stage consists of identifying a need for more, or a need to "do something" so that work, people and processes are not operating in silos.

Observe and Investigate in Stage 2

This is the stage where processes emerge. Work is happening, and some changes are taking place; but teams and departments still operate independently and don't really know about what each other does. This is also the stage where culture breeds perception as employees establish formal relationships with one another. This bond between culture and perception can make or break the progression of engagement to stage three: In the constructive scenario, employees believe change can and will be positive. Or the reverse could be true — they believe the organization won't invest in change and therefore any change will be seen negatively. Measurement in stage two should therefore focus on perception and innovation. You should be investigating what isn't working for employees, what frustrations they have, and what solutions they have for change.

Although this is the stage where processes emerge, some processes may just be in their infancy whereas others might point to the wrong solution altogether. Measuring perception of the emerged processes lets you know if you're on the right track.

Investigation should also focus on leadership. Understanding and creating a framework of your leaders' styles can help you measure engagement of subordinate employees and understand how style is impacting employee behaviors. A number of leadership questionnaires are available, such as the leadership skills inventory, the personal style indicator, and the leadership and capabilities model.

Use Analytics to Create a Baseline in Stage 3

This is the stage where collaboration is actively pursued. Ask employees to complete a mind map of all the relationships they have or are cultivating across the organization. Ask them to identify what channels and software they use to communicate, to access information, and to store and share their work. Ask employees if their relationships and work methods are currently working for them. Compare what they say against what you see in the analytics. Remember that you are measuring perception as part of a much broader organizational system. If employee feedback does not match the analytics, investigate why it doesn't match. If employee voice does match what you find in the analytics, discover if the process in question is the most effective way to do business.

Of course, not everything can be compared to analytics and not everything that goes on within an organization exists online. But what you essentially can do in stage three is use the analytics as a baseline for understanding how digital behavior changes over time and in different situations. If you find that employees are exhibiting behavior that is evident of stage four or five engagement, but you notice that analytics data has fundamentally changed, you can begin to explore why this change has occurred. You can then use that new data as a baseline for a different stage which may be related to a specific organizational event or time of year.

Employ Multi-Method Analysis in Stage 4

The jump between stages three and four is quite a big one. Stage four is when everything becomes integrated and everyone knows what's going on across work types. It is a highly functioning stage, and because it represents alignment of digital systems, it also means greater analytical insight.

In this stage you're not only examining what employees are looking at and what they're accessing, but you're also evaluating the content of what they are interacting with. What you're essentially looking for in stage four is accountability that the behaviors that are identified are actually relevant to the business vision. You can have characteristically engaged employees, but being able to attribute their engagement to the right business vision is a hurdle that must be crossed before stage five.

By using analytics and performing a content analysis, you can present the findings to the leadership group. This allows for a quality assurance check and provides greater insight into what areas might need improving before the business can progress.

Maintain Standards and Learn from Stage 5

This is the stage of engagement in which internal operations are all strategically aligned and the customer experience is streamlined. In level five, measurement is used to sustain and manage. When stage five behaviors are evident, you can use previous data to draw conclusions and delve deeper. For example, you can re-examine leadership questionnaire data, and start to look at what leadership styles are influencing proactive and desired employee behavior.

This is the stage where you can investigate the root causes of engagement and you can influence other areas of the business to achieve the same outcomes. This is also where you can compare the daily activities and routines of individual employees, and see how their location, work type and even demographic differences are contributing to an engaged workforce. For example, if you find that most employees are accessing content via a mobile device, you can begin to design future communication that caters to this preference.

In this stage, employees should have all the tools and working conditions they need to continue performing their job at the new high standard. However, though this is the final stage of maturity, it is never "finished." Instead, it is an ongoing process of fine tuning engagement. The organization must remain adaptive to change and responsive to employee feedback in order to maintain the business value and success achieved through this maturity process.

Conclusion

Although the five disciplines are all interrelated and should align with one another, some stages of maturity can progress before other areas. For example, it's much easier to improve communication practices than it is to transform the effectiveness of a particular leader. But what you may find is that even though some disciplines progress before others, they cannot continue to mature until the other disciplines have caught up.

Engagement can also be specific to work areas. Some departments can have stage-four behaviors with all five disciplines aligned, whereas other departments can still be operating in the earlier maturity stages.

It is also important to note that employee satisfaction should be measured across all five stages. This can give you valuable insight into what an employee wants and needs to be happy in their job, and can also expose problems with leadership and communication. Although satisfaction does not equal engagement, it can point to areas that are impacting the emotional well-being of employees.

Measurement can also include aspects of disengagement, such as absenteeism. Analytics of customer behavior and external communication should also be used internally to help with processes and procedures.

In closing, employee engagement is a constantly evolving practice that gains complexity as an organization matures. The more engaged employees become with their work, the more successful the organization will become with its clients. But the challenge is always going to be maintaining that optimal level of engagement. It is an ongoing investment, an art, a practice, an individual and an organizational commitment. You're not just measuring engagement, you're using the findings to change the organizational landscape and invest in improvements to those five main contributing disciplines: culture, leadership, communication, knowledge management and emotional intelligence.

Further Reading

Attridge, Mark. "Measuring And Managing Employee Work Engagement: A Review Of The Research And Organisation Literature." Journal Of Workplace Behavioral Health 24.4 (2009).

Mishra, Karen, Lois Boynton, and Aneil Mishra. "Driving Employee Engagement: The Expanded Role Of Internal Communications." Journal of Business Communication (2014).

Shuck, Michael Bradley, and Karen K. Wollard. "A Historical Perspective of Employee Engagement: An Emerging Definition." Eighth Annual College of Education & GSN Research Conference (2009).

Sunder, Vijaya. "Six Sigma – A Strategy For Increasing Employee Engagement." Journal For Quality & Participation 36.2 (2013).

Steffens, Niklas K., et al. "Leaders Enhance Group Members' Work Engagement And Reduce Their Burnout By Crafting Social Identity." Zeitschrift Für Personalforschung 28.1/2 (2014)

Zafft, Carmen R., Stephanie G. Adams, and Gina S. Matkin. "Measuring Leadership in Self-Managed Teams Using the Competing Values Framework." *Journal of Engineering Education* 98, no. 3 (2009)

ABOUT THE AUTHOR

Natalie Hardwicke currently works as a freelance writer and communication consultant, and has more than five years' experience working in communication and employee engagement managerial and advisory roles in the Australian Public Sector. She is in the final year of her Masters in Publishing and Communications at the University of Melbourne, with her thesis looking at how enterprise social media is changing internal workplace communication. She holds a Postgraduate Certificate in Arts (Editing), also from Melbourne, as well as a Bachelor of Communication (Honours) and a Bachelor of Science in Psychology from the University of Canberra. Natalie provides advice on the "human aspect of business communication" on her professional blog, speakingofcomms.com.

Chapter 1

The Economics of SharePoint and its Social Implications

by Jeff Shuey, Chief Evangelist, K2

SharePoint took a long and relatively slow ride to the top of the heap in the Enterprise Content Management (ECM) space. SharePoint as a product is going on 15 years of age now. It's an adolescent in the market, and with the release of SharePoint 2013 is on version 5 of the Microsoft SharePoint product evolution.

For those who haven't seen it yet, there will be a SharePoint 2016. This will likely be the last on-premises version of SharePoint. It's been a good ride and one that has secured Microsoft and the partner ecosystem that support it billions of dollars in revenue for software and services.

And it's not over. There is a whole new world opening up with the recent focus on the App Model and the rise of cloud computing. The Microsoft ecosystem and the people that have grown their business on SharePoint — both as customers and partners — have a few years to get these new paradigms under their belt...before the next wave hits.

A Little SharePoint History

The rise and deep market adoption of Microsoft SharePoint has been nothing but remarkable.

83% of the Fortune 500 firms use SharePoint.

I remember when I first heard about what was to become Microsoft Share-Point. Back then I was working for FileNet, which at the time was one of the Big Three of the ECM industry. More on that later. In the early days SharePoint was code-named "Tahoe" and there were grand ambitions by the leader of the team. They were going to come after the Big Players. They saw a gap and an opportunity.

The Big Three ECM Players

FileNet, Documentum and OpenText were the Big Dogs of the Enterprise Content Management space back in the '90s and into the early 2000s. However, they had the same challenges most every other legacy vendor faced. It's hard to stay ahead of the curve.

SharePoint sought to make a course correction for the Big Three. And they did. Jeff Teper, Corporate Vice President at Microsoft, is considered the father of SharePoint. He has done a masterful job guiding SharePoint from the early days of a relatively weak product to the powerhouse it has become.

SharePoint has become a lot of things. Most companies do not take advantage of all of the features SharePoint has to offer. This is one of the factors that makes an economic analysis difficult, at least from the utility economics perspective. Because there are features within SharePoint that can help almost any business. Yet they go unused, sometimes because other systems are in place. Other times because there may not be a perceived need for these functions. And sometimes it's because the people running the SharePoint deployments either don't know about the features or they don't have time to adequately deploy them.

That's all starting to change with the App Model — where end users, line of business managers and even executives are taking the no-code/low-code option of Bring Your Own Applications (BYOA) to the forefront.

What Were The Drivers That Made SharePoint Possible?

Like a lot of legends . . . SharePoint was in the right place at the right time.

But this is not a love letter to SharePoint. There were and are fundamental drivers of economics that allowed SharePoint to succeed where others have either failed completely or failed to maintain their success over time.

The economic drivers of that have helped make SharePoint successful are marginal utility and consumption economics. From a marginal utility to a consumption economics model, there is some rhyme and reason for the success of SharePoint to date. This is not to say the early team that drove the product was not influential. They were. They did a masterful job of getting the product out there.

Definition and Use Case for Marginal Utility:
In economics, marginal utility can be applied the additional satisfaction or benefit (utility) that a customer derives from buying an additional unit of a product. In the SharePoint world the first sale is often hard fought and a challenge. The second sale into the same account is often much easier and the perceived value of the solutions SharePoint is being used to address may also be perceived as being of lower value. This allowed SharePoint to penetrate more deeply into customer accounts because the barrier to the next sale was reduced.

As with all things there are exceptions. Some of the assumptions made with marginal utility are that the SharePoint deployments were successful. Which means the solutions were adopted and become at least somewhat mission critical, at least in the sense that they were fulfilling an important role within the organization. When this happened it was often easier to make the next sale into the same account. Which often made it easier to sell into other accounts in the same industry by utilizing the power of recommendations from satisfied customers.

Definition of Use Case for Consumption Economics:
Simply put this is when the *User Drives the Tech Decision.* Consider the early Bring Your Own Device (BYOD) scenarios of the past

5 years. Employees brought their own devices, predominantly iPhones, to work and expected the IT department to adapt. If the next 5 years this will continue with a Bring Your Own Application (BYOA) model. Where the end user will expect to use applications they are familiar with and, again, they will expect IT to adapt.

Out There Does Not Equal Adoption

As some pundits have noted, a little over 80 percent of the Fortune 500 companies own SharePoint. But ownership does not equal adoption. SharePoint had to struggle to get companies to use SharePoint for something more than a document repository.

The Fastest Growing Business in Microsoft History

This is what Jeff Teper used to say about SharePoint. Today this statement applies to Microsoft Office 365. This is not surprising. Nothing lasts forever. One of the drivers of economic success is the ability to adapt to market demands. While SharePoint has adapted with SharePoint Online, which is a component within the Office 365 brand, there is still a sizeable market that continue to use SharePoint on-premises.

Law of Large Numbers

At some point it could be argued that SharePoint was impacted by the Law of Large Numbers. However, I don't think that is the case. I think it's just a change in the way people want to do their work (think mobile) and the way companies are allowing and expecting platform vendors to react (think cloud).

> Definition of the Law of Large Numbers:
> It is a principle of probability according to which the frequencies of events with the same likelihood of occurrence even out, given enough trials or instances.

SharePoint vs. Office 365 Adoption and Growth Trends

SharePoint is a fixture, but it is being impacted by its former baby brother, Office 365. Office 365 is growing up and spreading its proverbial wings. That's not to say that SharePoint On Premises (OnPrem) is going away any time soon. The market will drive behavior here. As we have already seen Microsoft has announced another OnPrem version of SharePoint. The economic indicators are that OnPrem SharePoint 2010 is trending down, but it will be a long time before that dips into single digits. While Share-Point 2013 is experiencing over 100% Year-over-Year growth. This growth rate will taper off as more and more companies, especially in previously untapped markets, begin to adopt Office 365.

Office 365 will definitely continue to gain market share. These gains in market share exist for many reasons (see Table 1). One significant factor is that Office 365 will continue to add more and more features that will never be available for the on-premises versions of SharePoint. The economics of a cloud based deployment continue to be easier to justify for a lot of companies. Office 365 makes it very easy to get started while putting the burden of uptime, backups, security patching and so much more on Micro-soft. As Microsoft has shown they are quite adept and providing a secure and reliable productivity platform.

Interestingly, there is a partner play here. I mentioned that Office 365 will continue to add features faster than SharePoint OnPrem. This is where partners can and will come in. The SharePoint partner ecosystem will always be on the lookout for gaps in SharePoint product capabilities and seek to fill them with solutions. This is why I say partners and community are the secret sauce for SharePoint. Partners will seek gaps in SharePoint OnPrem, Online and in Hybrid scenarios. If there is a market there ... the partners will seek to fill it. For this reason I expect the percentages of SharePoint and Office 365 adoption to be somewhat volatile as customers make decisions on the best model for their business.

Table 1. SharePoint and Office 365 Adoption

Percentage of business users using Office 365 and SharePoint 2010/2013			
Product / Year	**2012**	**2013**	**2014**
Office 365	12	15	23
SharePoint 2010	81	79	71
SharePoint 2013	7	16	35

Source: Forrester Global IT Usage Survey

Update: Yes, Virginia, There Will Be a SharePoint 2016

In the SharePoint 2016 release, you will see some features coming to on-premises SharePoint, according to Julia White's blog post on the *Evolution of SharePoint*. But the pace of innovation, combined with demands of the customer community, will necessitate a shift to cloud-based offerings.

The fact is that some companies still prefer to keep their servers on-premises and under their control. Whether or not this is a wise choice is irrelevant. If the company wants to incur the costs of managing servers, doing backups, adding security and maintenance patches and generally having the satisfaction of being able to walk into their server room and point at the place where they think their SharePoint Servers are running, then so be it. More power to them.

The reality is that some companies, notably government organizations and those that support them, feel this is the most secure way to manage their data. They might be right. Whether or not they are right, it's their prerogative to decide that they want to run an On-Premises SharePoint Environment (OPSE). There are a lot of people, partners and professionals, out there that will rally around the idea of an OPSE. The good news is that many of the best practices, policies and procedures are already in place.

One area that I suspect will continue to be a challenge with On-Premises SharePoint Environments is when they need to connect with other systems that are hosted in the cloud. Of course, they are doing this today and nothing really changes. I suspect that tighter controls will be created to monitor the connections between systems, as well as the flow of data between OPSEs and cloud-based solutions.

However, I do think that more and more, these firms will be in the minority. Partners to the rescue! Do not be surprised when a partner ecosystem springs up to support those that want to stay on-premises. This is one of the best-kept secrets of the Microsoft Partner Ecosystem. The Microsoft Partner Ecosystem is one of the best in the world. One of the things Microsoft partners have figured out a long time ago is that there are markets for almost everything. And these partners strive to fill them. And Microsoft will strive to make a competency within the space. The circle of life continues.

This leads us to the hybrid solutions.

Hybrid Is Real

As of last year, Microsoft began allowing us to use the "H-word" — *hybrid*.

For a long time the powers that be at Microsoft had poo-pooed the idea of hybrid computing. That's all changed. It could be the new Microsoft CEO, Satya Nadella, with his *Mobile First — Cloud First* mantra. Or it could just be a realization that the market is demanding hybrid offerings.

BYOA Opportunity & the Partner Opportunity

As I mentioned, a new wave is rising. *Bring Your Own Applications* is just taking off. The good news is that the majority of the hard work and the heavy lifting has already been done. Phase one was the rise of Bring Your Own Device (BYOD). IT professionals, organizations and the platform vendors that support them scrambled to accommodate people bringing their iPhones, iPads and other "nonstandard" devices to work with them. IT and platform vendors stepped up and did a very good job making it possible to bring the device of your choice into the corporate environment while maintaining security, governance and compliance.

The early BYOD efforts have paved the way for phase two: The BYOA phase. Employees are going to start bringing the applications of their choice into the corporate environment. Some people are already used to doing this with third-party content storage sites, such as Dropbox and Box, as well as with digital signature options and multiple CRM offerings.

The good news is that Microsoft has firmly embraced this. Recent advances include releases of Microsoft Office for iOS and Android, as well as the recent adoption of Dropbox, Box, Citrix and Salesforce[1] as first-class citizens within the newly announced Microsoft Cloud Storage Partner Program.[2]

The App Economy

This could be described as Sharing vs. SharePoint. The past 10 years have seen a dramatic rise in the sale of smartphones. This segment has traditionally been driven by the Apple iPhone, but in the past few years the iPhone's lead has been replaced by Android-powered devices.

1.2 billion smartphones were sold in 2014. That's one for every seven people on the planet. *Source: Techcrunch* [3]

17

The App Economy is driven by the use of smartphones for everything from catching up on email and social media feeds to more business-specific functions like signing a document or approving an expense claim.

The consumer economics aspects of the App Economy are directly related to the rise of mobile computing and the demands from customers that solutions be made available on their devices. This, in turn, has led companies to cater to their employees and create solutions for supporting them. Of course, the companies need to insure security, governance and compliance risks are all managed.

Most employees could not care less about the challenges the IT organization as a whole faces. And when they get bored with an app, they are likely to look for an alternative. This leads to the consumer economics Rule of Diminishing Marginal Utility, when the employee has seen all they think they need to with the app in question and move on to the next one. And this reinforces the whole BYOA model and mentality. The point here is that companies will need to continue to work hard to stay ahead of the curve — or at least to try and keep up with the trends.[4]

The Opportunity Cost of SharePoint

It's tough to argue with free. SharePoint was offered at no cost to a lot of customers through a Microsoft CoreCAL licensing arrangement. This drove a lot of "seat counts" and "logo wins" in the early days of Share-Point — which, by the way, was a brilliant move by the SharePoint team to drive the product into accounts. Most products don't have this luxury and the SharePoint team was wise to take advantage of this unique opportunity.

The challenge of opportunity cost is: *What could have been done with those resources?*

The brilliant part of this is that instead of taking time to buy licenses, many companies had both skunkworks and actual projects where they were testing, trying and deploying solutions based on their SharePoint licenses. Of course, eventually those free licenses were often upgraded to include higher-level capabilities, which is where the billions of dollars in software and services revenues came from.

What else could those funds have been used for? Well, many times they were allocated to other ECM systems. Often these were legacy systems that were slowly being phased out. As SharePoint continued to improve from version to version, these legacy systems became less relevant and were put out to pasture. This, again, was a brilliant move by the SharePoint team. They created core capabilities for base-level functionality related to records management, document management, business intelligence and to a lesser extent forms and workflow capabilities. They left wide margins for partners to fill these gaps. And this enabled a large partner ecosystem to thrive and flourish.

This was the genius of the SharePoint economics. There is a sizable contingent of SharePoint fanatics that have built their careers around the Microsoft Partner Ecosystem, myself included, and specifically around the SharePoint product offerings.

Why SharePoint Succeeded

It's all about the ecosystem. Microsoft has built and fostered an incredible ecosystem of partners.

There is a large group of partners that have created solutions and services that live inside the SharePoint product offerings or that operate somewhat autonomously outside of SharePoint. Some of these external features and capabilities are very broad and horizontal in nature — features like document scanning, fax services, backup and restore functions, etc. See the ECM Pillars graphic below.

Other features are more specific to SharePoint — partners have extended or enhanced built-in functionality like site migration, the use of document types and content types, and workflow to provide a richer experience

In addition there is a core group of partners that are what Microsoft calls Most Valued Professionals (MVPs). The MVP community has made it possible for Microsoft to provide an additional layer of support for both partners and customers. The MVP program is an incredibly model of economic efficiency in that Microsoft spends very little to get a very sizable return on their investment.

MVPs –This is an exclusive club of non-Microsoft employees that have taken their time to become experts in various Microsoft products, technologies and services.

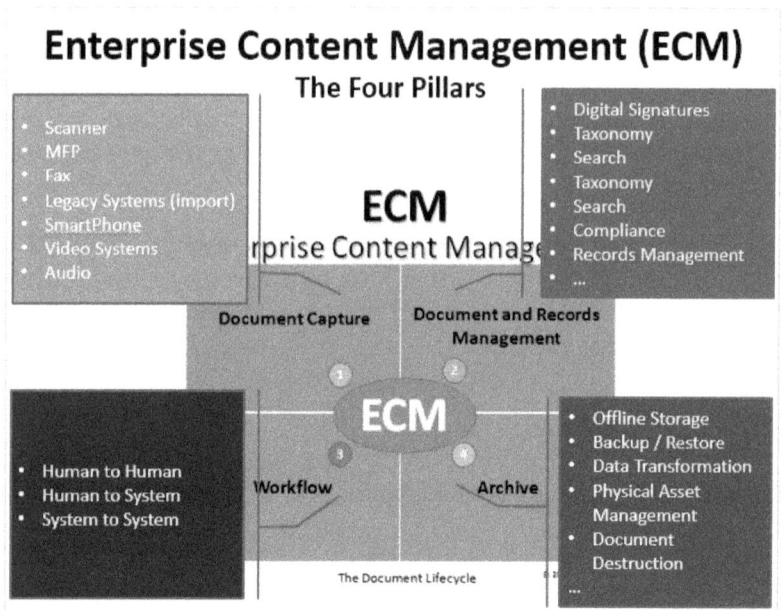

Enterprise Content Management (ECM)
The Four Pillars

- Scanner
- MFP
- Fax
- Legacy Systems (import)
- SmartPhone
- Video Systems
- Audio

ECM
rprise Content Manage

- Digital Signatures
- Taxonomy
- Search
- Taxonomy
- Search
- Compliance
- Records Management
- ...

Document Capture

Document and Records Management

ECM

- Human to Human
- Human to System
- System to System

Workflow

Archive

- Offline Storage
- Backup / Restore
- Data Transformation
- Physical Asset Management
- Document Destruction
- ...

The Document Lifecycle

Community Matters

The biggest factor that has led to the economic success of SharePoint is the community. There is a coalition of avid fans that support, defend and engage globally to share their successes, failures and best practices.

The "secret sauce" of SharePoint is the community. Yes, this might seem like I'm drinking too much Kool-Aid, but the fact is that there is a globally distributed group of SharePoint enthusiasts that will jump at the chance to help a fellow SharePoint enthusiast.

Social SharePoint

There are groups dedicated to supporting the SharePoint machine. These, by default, include the direct and indirect partner solutions and services that drive SharePoint adoption and success. These groups can be found online and in person. There is a Twitter group that responds to the hashtag #SPHelp. There are countless SharePoint User Groups (SPUGs) in cities all around the world. There is a great all-volunteer organization called SPSevents.org that helps coordinate SharePoint Saturday around the globe, where they get people to spend their time to come to events and share what they know about SharePoint.

Community matters! This might be the most important ingredient to the economic success of SharePoint.

Conclusion: The Economics and Future of SharePoint

The future of SharePoint is bright-ish. I don't see SharePoint itself disappearing any time soon. However, I do see more and more morphing into a cloud-first model. This should not come as a surprise to anyone who has been watching or living in the computing world in the past few years. The economics show that the world is shifting to empower the user with code-free, easy-to-configure, mobile-ready solutions.

It is almost guaranteed to be a low-code or completely code-free model that enables anyone in the organization to create solutions that exactly fit their needs. No longer will end users or executives with an idea be bound waiting for the IT organization to get something on their calendar and deliver it when they can.

Early Spreadsheets Are Bellwethers for the Future of SharePoint

Those old enough to remember the power of Dan Bricklin's VisiCalc and Mitch Kapor's Lotus 1-2-3 will know the feeling. Those a little younger may remember the first time they created a macro or ran a huge list of numbers in Microsoft Excel. Early spreadsheet users and the solutions they created changed the way we see, use and process data.

The SharePoint of the Future will be similar. The users are in command.

NOTES
[1] *New cloud storage integration for Office - http://blogs.office.com/2015/02/17/new-cloud-storage-integration-office/*

[2] *Cloud Storage Partner Program - http://dev.office.com/programs/officecloudstorage*

[3] *1.2B Smartphones Sold In 2014 - http://techcrunch.com/2015/02/16/1-2b-smartphones-sold-in-2014-led-by-larger-screens-and-latin-america/*

[4] *5 TIPS TO THRIVE IN A BYOA WORLD - http://jeylabsblog.com/2015/02/15/5-tips-to-thrive-in-a-byoa-world/*

ABOUT THE AUTHOR

Jeff Shuey is an expert in the Enterprise Content Management industry. He brings over 20 years of Channel Sales, Partner Marketing and Alliance expertise to audiences around the world in speaking engagements and via his writing. He has worked for Microsoft, Kodak, and is currently the Chief Evangelist at K2 (www.k2.com). Tweet him @jshuey (twitter.com/jshuey) or connect on LinkedIn (www.linkedin.com/in/jeffshuey), Facebook (facebook.com/jeff.shuey), or Google+ (gplus.to/jshuey). He is active in the Microsoft Partner Community and is the co-founder and President of the IAMCP (www.iamcp-us.org/?) Seattle chapter.

Jeff is a contributing author to Entrepreneur (www.entrepreneur.com/author/jeff-shuey), Elite Daily (elitedaily.com/money/create-linkedin-profile-actually-gets-noticed/), Yahoo (smallbusiness.yahoo.com/advisor/blogs/profit-minded/smarts-luck-hard-222559522.html), US News (money.usnews.com/money/blogs/outside-voices-careers/2014/03/27/planning-the-perfect-working-vacation?utm_source=westconsocialondemand.com&utm_medium=Westcon+-Convergence&utm_campaign=westconsocialondemand.com) and to the Personal Branding Blog (www.personalbrandingblog.com/author/jeff-shuey/).

Chapter 2

How to Define Your Intranet Metrics to Increase Adoption

by Kanwal Khipple

Organizations, including yours, invest in an intranet for a variety of reasons. To fulfill an organization's objectives, an effective intranet must serve one or more of the five functions of a modern intranet:

- Documents portal
- Corporate communications portal
- Business process management portal
- Team collaboration portal
- Social hub

Your portal's success lies in how well it enables employees to get things done. Is your intranet an unwieldy and marginal application that employees use because there is no alternative and only a fraction uses it on a daily basis? Or is your intranet an essential application such that employees can't imagine their day-to-day work without the tools and resources it provides?

The value proposition for intranets depends on implementation and how critical the application is to your organization. How can you measure your success in achieving that? Is your intranet visited by employees every day because it performs well, or only because the portal has been set as the employee homepage? Are employees spending a lot of time on your intranet because they find it a valuable tool, or are they staying longer because they have a hard time finding valuable information?

Defining what metrics you want to measure is critical to understand not whether your portal is successful or not. Be careful of the metrics you capture as they can have impact on your portal and its intended purpose. There's been numerous occasions where organizations have used the wrong metrics and steered their portal in the wrong direction. Impacting the target audience to the point of confusion. If you don't know what you are measuring, metrics are irrelevant.

The good news is that defining effective metrics is not difficult, and that's exactly what this chapter covers. This chapter focuses on helping you identify the type of portal that best meets the needs of your organization and employees, and on helping you define appropriate metrics. I will also discuss some of the leading metrics that you can use to measure your portal's success. My goal is to enable you to create an ongoing adoption plan to improve your metrics and continue your journey.

Core Benefits of Your Intranet

Is your portal valuable? Depending on whom you ask in your organization, you can find opinions that vary across a wide spectrum: a small percentage of people love the intranet, a small percentage hate it and the vast majority don't really care. This is very common in small- to medium-size organizations, because employees often don't understand the ROI for the intranet, and more importantly don't see how it can provide them value.

Your company could be early in its intranet maturity and have a very basic document management portal, or it could be further along its journey and have a social intranet. There are many factors that can determine the goals of your company intranet. Where your company is in terms of technology adoption maturity, corporate culture and company strategy are all factors to consider when planning the goals of your intranet.

What follows is usually the toughest question: If I'm going to invest in creating Intranet v2.0, how can I ensure it provides a higher ROI than v1.0? Your metrics will help you determine the value of your intranet. Before measuring the value, you first need to define the company's goals. Let's start with the basics to get a better understanding of your portal and what you hope to achieve. Table 1 lists the five main functions of a company intranet, with common activities they support and company goals they promote.

Table 1. Five major types of intranets. Which one is yours?

Portal Type	Common Activities	Company Goals
Documents Portal	• Providing central databases for document posting and exchange • Posting, editing, and updating documents efficiently and traceably	• Manage document development and exchange centrally • Maintain strict version control and tracking
Communications Portal	• Promoting corporate culture and communicating brand messages • Presenting information interactively to encourage engagement, idea and knowledge exchange, and sharing • Publicizing updates, announcements, and new product or process rollouts	• Provide an enterprise-wide communications hub • Broadcast company messages and share and distribute content (one-to-many communication) • Enable multiple users to contribute and receive information (many-to-many communication) • Promote culture • Encourage employee engagement and information exchange • Strengthen the brand internally; truly become "one company"
Business Process Management Portal	• Providing central repositories for standards, policies, workflows and process guidelines • Automating and standardizing document creation with templates • Creating alerts to announce changes and new additions to the intranet	• Standardize processes • Automate workflows • Centralize tasks • Increase efficiency, eliminate duplicate efforts

Portal Type	Common Activities	Company Goals
Collaboration Portal	• Working collaboratively within and across business units (for example, multiple people working on one document) • Enterprise-wide searching for employee data, such as contact information, specialty areas, group membership, and personal interests • Establishing working relationships for collaborative support and sharing of knowledge and expertise	• Collaborate better • Search across departments • Find relevant information • Facilitate work in groups • Share knowledge • Search for people and expertise • Modernize digital tools and work methods
Social Portal	• Providing timely information to employees, customers and business partners • Encouraging employee engagement and ongoing interaction	• Innovate through ideation exercises • Attract talent • Share knowledge better • Build a strong employee community • Strengthen the brand internally; truly become "one company" • Increase employee engagement • Search for subject matter experts

Use this list to help you assess and develop a strategic intranet plan for your organization. Identify what are the objectives that you are achieving with your intranet. Next, let's identify specific business problems and processes that are linked to company goals, and could be improved with targeted use of the intranet.

Deriving Value from your Intranet

Understanding who your target audience is one of the most critical stages of any project. Not only does this drives the core benefits, as we outlined earlier, but also drives what your target users are hoping to accomplish on a day to day basis. As you look at your audience groups, you'll want to select the right mix of recipients to capture requirements.

The traditional approach, many organizations follow, is to have Management/IT drive the requirements the target audience will need and implement the portal. My suggestion and recommendation to you is to start by doing some research and talk to people within some key departments (examples: HR, Finance and elsewhere) to identify the challenges to achieving a goal. Following this initial discussion, engage with users who are able to represent the target audience groups and rather additional requirements as a part of working sessions.

Far too often, these discussions focus on how the solution should function from a technical perspective rather than on what the current challenges are from a business or cultural perspective. Focusing on the challenges the teams are facing allows you to target the core issues and assess whether a simple solution already exists. This simplifies the approach and enables the team to visualize the Minimum Viable Product (MVP).

Let's say our company wants to achieve the following goal: "Improve the discovery of subject matter experts by sharing knowledge and promoting expertise."

If employees tell you that they have no central place to post ideas and ask questions, your solution could include rich community sites with discussion forums, blogs and knowledge-based wikis. In this example you can link the benefits of a social intranet to a specific business problem to be solved, and to achieving the larger company goal. Employees also demand that they need access to the community wherever they are and with whatever device they have. You'll also find that personalization is at the top of most requirement lists. These working sessions will surely give you dozens of requirements.

This is just one example. During your discussions, look for specific examples of how technology can benefit the culture and businesses processes to increase productivity. Once you've found several examples that support company goals, you're ready to write your intranet strategy.

Compile a list of the benefits you are looking to achieve from your intranet, and from there you can derive a list of the technical features necessary to get you there. This enables you to focus on the MVP for each feature and drive out an intuitive UX and priority-driven delivery schedule. Table 2 is an example of a company's outline of benefit requirements and the directly related features.

Table 2. Examples of Intranet benefits and features

Benefit	Related Features
Support for ideation and innovation	• Rich, discoverable profiles • Search-driven communities of practice • Q&A forums for key business topics • Microblogs and status updates • Hashtags to bring related communities together • User-generated blogs • Group-generated reference material
Greater admin efficiency	• Purposeful, task-oriented content • Effective, relevant search functionality • Simple and intuitive navigation • Single sign-on • Integrated applications
Keeping employees in the loop	• Personalized news • Corporate and topic-based news • Visually appealing news • Ability to share and like news
Streamlined collaborative procedures	• Centralized team collaboration sites • Reduced use of email for document management • Multi-user editing • Group/iterative editing of shared files • Online discussions captured centrally
Increased employee satisfaction	• Increased employee ability to speak up and be heard • Ability to make meaningful contributions • Connections with colleagues • A feeling of connection to larger community
Attraction of talented new staff	• Modern digital working environment • Employee ability to speak up and be heard • Obvious opportunities to contribute
Stronger employee bonds	• Employee-generated news and content • News about employees • Personal/fun online communities • Commenting features throughout the site • Profiles with tags for shared interests • Open, discoverable groups

Having this clear mapping of benefits to features not only gives you a clear idea of your intranet strategy, but it also opens the door for discussions on assessing your portal's current performance as a baseline and keeping track of your progress.

The key to understanding your intranet's core benefits is to discuss the current challenges and see how you can pivot to deliver value. There are times when you'll notice that the technology alone does not have the biggest impact on your ROI, but rather a combination of technology, processes and culture.

How to Define Metrics

Let's imagine you have a vague communication goal to "make people love the company" and you work with the marketing department to create and publish an inspirational video. You can feature the video on the intranet homepage and only a quarter of the company clicks on the video to view it The reaction of the audience that did see the campaign may span a range from positive ("Wow!"), through neutral ("It's alright."), to negative ("This was a waste of my time!"). Basic implementations usually don't provide metrics on whether a video was watched or not – only if users hit the page – which means you have an even greater need to include like / comment functionality to measure reach. If you've created a compelling and inspirational video, you'll notice the following:

- A minority will demonstrate their reaction with a like, a share or a comment.
- Whereas the majority will view the video, close the browser window, get on with their day and rapidly forget about it.

In order to gauge whether the video had impact or not, you could view the number of comments as a measure that the video was successful. However, this doesn't accurately reflect the impact of the video on engagement. Why not? For one, comments can be both positive *and* negative, so a strict count of these to demonstrate "engagement" is pointless. If people share or like a post, that means they would recommend it and the video had an impact on them. Negative reactions can also be captured by looking at bounce rates and, in the case of video, how soon viewers abandoned the video after it started playing. Very few organizations invest in this. Look at other aspects of intranet engagement such as average time on page. If you were trying to understand whether last week's featured video was as successful as this week's, you might compare average time on page between both time periods. Ultimately, being able to measure anything is fairly pointless unless you know why you are doing so.

Our objective is to understand, measure and track each benefit.

If you have a communication goal as vague as "making people feel great about the company," it should hardly come as a surprise that this is difficult to measure. However, consider a communication goal like "Make sure everyone has filled in their quarterly performance review by March 31st." This is directly measureable, and metrics are less about page views of the campaign and more about the proportion of reviews filled in on time to those that are not. The metrics we choose clearly communicate what we are hoping to prove. Another recommendation to remember: Use measurement to figure out what you are going to do better next time, rather than as a way of justifying how great the portal is today.

Define metrics to focus on improving your intranet.

One way you can help improve your engagement levels is to communicate how amazing your portal is. Perhaps when displaying that inspirational video, you also include a tool that displays the number of people who liked or shared the video? Not only does everyone like to see that what they have posted on the intranet is popular, but everyone also likes to view content that is popular and liked by others. The advantage of displaying content popularity is that it encourages more people to share, more people to click, more people to discuss, and perhaps more people to continue coming back to the intranet. One other recommendation: don't underestimate how much impact design has on engagement.

Top Metrics That Ensure Success

After providing you some insight into how and what should be measured, I want to start you off with a few strategic metrics for your intranet. These are some of the most popular metrics that we leverage in our engagements for customers. Time and time again, customers see value in defining a baseline and keeping track of these metrics.

This list goes beyond outputs such as page views or time per visit. While those are definitely useful metrics to monitor, they don't get to the heart of the matter. The outcomes, or rather, the results of intranet activity are the most important. These metrics are much harder to generate and report, but they provide insight into your intranet strategy that can help clarify the ROI and gauge whether you are achieving the purpose of your intranet.

1. Intranet Adoption Rates

Metric focus: Users who participate versus users who do not participate on the intranet over a given period of time.

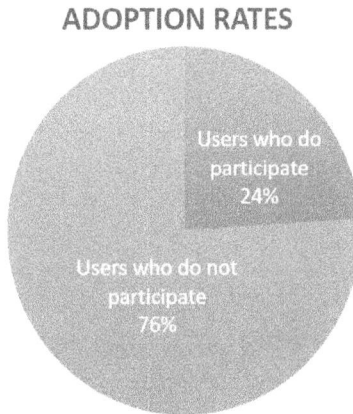

ADOPTION RATES

Users who do participate
24%

Users who do not participate
76%

What this measures: The number of users who actively participate (create/edit pages, make comments, interact with others in communities, etc.), presented as a percentage of total users.

How to read it: When viewed over time, this metric tells you whether or not people are using the intranet for more than reading news. Are they actually contributing to the organic growth of your intranet?

Target audience: Your core intranet stakeholders, departments such as HR, IT and Internal Communications.

What's next? Depending on your platform, this isn't an easy metric to capture. If your data is changing drastically for some time, you might also want to conduct user surveys or collect additional data. Capturing this feedback will guide you to provide more training, produce more relevant and interesting content, engage more actively with community managers, or develop highly targeted and useful ways of using the interactive features of your social intranet. I would recommend to be prepared that if you are able to produce this metric for the organization, department or functional area managers are going to want breakdowns for their own teams – e.g. what is our participation level in Research vs. in Sales? Same goes for geographical locations – what is our engagement in the New York office vs. San Francisco? People will want to slice and dice this information for many reasons.

2. Intranet Champions

Metric focus: The most active users of the intranet.

What this measures: The top contributors for the site—the top 20% who represent 80% of site activity.

How to read it: Who are your intranet champions? As a rule of thumb, 80% of content is created by 20% or less of the users. Ensure that you always capture feedback from these leaders. By strategic interaction and collaboration with these folks, you can change your entire intranet quite quickly.

Employee	New Pages Created	Pages Edited	Comments Contributed	Total Actions	Percentage of Total Actions
Jessica Mitch	42	283	59	384	24%
Harsimran Kaur	33	167	44	244	15%
Ronald Quinsten	25	142	41	208	13%
John Mitch	31	155	36	222	14%
Arman DeSouza	16	67	39	122	8%
Prasad Haseeni	14	36	50	100	6%
Dave Matthews	14	28	17	59	4%
Pedro Sanchez	12	33	23	68	4%
Tony Lopez	9	10	27	46	3%
Paras Smith	10	14	22	46	3%
All other employees	15	32	67	114	7%

Target audience: Your intranet team.

What's next? View these people as leaders. Connect with them frequently. They will guide and assist you in understanding the employees and how to get them to adopt your intranet. The champions can be your biggest allies in promoting use of the intranet. Find ongoing ways to recognize your intranet champions for their contributions.

3. Top 5 News Articles

Metric focus: The top 5-10 news articles on the intranet over a given time period.

What this measures: The news stories that had the highest rates of views and comments over a given period of time. For the majority of intranet portals, this metric is sufficient. For larger organizations, you'll want to look at the top 5 news articles in several departments or categories.

How to read it: Knowing the types of news stories that employees like the most and that drive the most traffic can help you understand employee interests and what they are looking for from your intranet.

Target audience: HR, Corporate Communications and any other department that publishes a lot of news.

What's next? Keep an eye on what stories, themes and topics resonate with employees. It'll help you get a better understanding of what employees find interesting. Sharing this information with employees as well as providing it to your intranet team will guide you in shaping the kind of intranet your employees are expecting.

4. Popular Keywords

Metric focus: The top keywords in employee searches.

What this measures: The most frequently used terms employees enter in the search box, over given periods of time.

How to read it: Knowing what people look for helps you understand their interests and how they search for information.

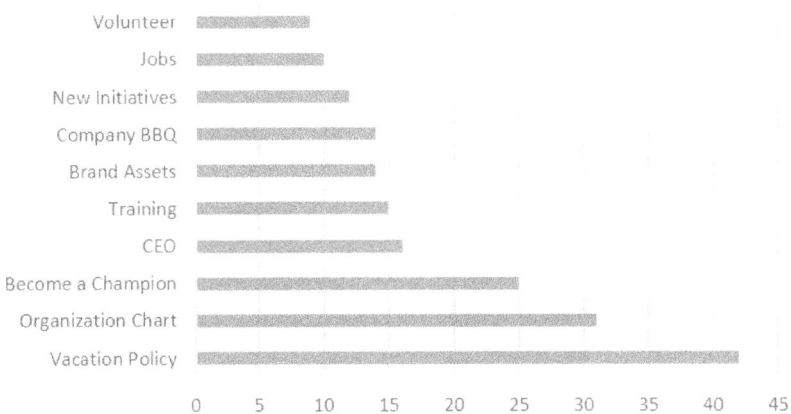

Target audience: The intranet team and specific departments who are responsible for sought-after content.

What's next? Once you have your list, try searching for these terms yourself and see what the top 3 search results are. Ensure that employees are getting the right information in their searches. If content is missing or the wrong results are being returned, this should be a key part of the process to ensure that the appropriate functional area provide the content.

Measure for Success

An intranet portal helps employees gain knowledge and make better and more informed decisions. It also helps reduce costs, saves time, increases collaboration, and promotes productivity and effectiveness. Your intranet portal can help employees find information more easily and perform their jobs better. Few portal designs are optimal right out of the box and require tweaking and customization. In smaller companies, designers can realize some features found in off-the-shelf portal software through simpler (do-it-yourself) means.

Most intranets have become completely unwieldy and present a highly fragmented and confusing user experience, with no consistency and little navigational support. With a strategically designed portal you can correct this problem by presenting a single gateway to all corporate information and services. One benefit of creating this consistent look and feel is that users require less time to learn how to use the environment. They also more easily recognize where they are in the portal and where they can go — no small feat when navigating a large information space. By integrating services and presenting personalized snippets on the initial screen, intranet portals also reduce the need for users to browse far and wide to obtain needed information, thus making it easier for them to perform their jobs.

Though measurement is an important tool in developing an effective intranet, measurement is only a small part of the varied aspects of a successful portal. Use the tips I've provided to help you create an ongoing adoption and improvement plan, improve your metrics and continue your journey to the perfect portal.

Do these metrics help you measure the success of your intranet? This chapter's author, Kanwal Khipple (kanwal@2toLead.com), would love to continue the conversation and learn how your organization is measuring success.

Chapter 2

About 2toLead

2toLEAD

You'll love the way we work.
Together.

2toLead is a generous consulting services company with a unique thought leadership and innovative thinking. We focus on Business Strategy, User Experience and Technology expertise to help transform organizations, small & medium business as well as startups. Our approach is centered on working in partnership with you to deliver results.

You'll love the way we work. **Together.**

Our team includes award-winning visionaries, published authors, designers, technologists and strategists. Our culture is really important to us and we feature an energetic environment where people strive to create the best work of their lives.

For more information, visit our website at www.2toLead.com, like us on Facebook and follow us on Twitter @2toLead.

ABOUT THE AUTHOR

Kanwal Khipple, Founder & CEO of 2toLead, is a leading User Experience expert within the SharePoint industry, with experience in building award winning portals and solutions that take advantage of Microsoft's Cloud platform (SharePoint, Office 365 & Azure). Kanwal's drive for success as the Creative and Technical Lead on projects has garnered him as a recipient of the Neilson Norman award for Top 10 Intranets (2014 & 2015).

Kanwal's passion lies in continuing to push for user experience innovation when redesigning intranets for majority of the largest brands in the world. He continues to preach on the importance of designing with usability as the primary focus. Kanwal's thirst to share knowledge has made him a prominent figure within the SharePoint community. Because of his passion and his involvement in many community driven events including launching successful user groups in Canada and the USA, Kanwal has been recognized as a SharePoint MVP by Microsoft (2009 to 2013) and most recently as an Office 365 MVP (2014-2015). He's also co-authored a book on Pro SharePoint 2013 and Responsive Web Development http://amzn.to/sp2013rwd

Feel free to reach out to him if you'd like to discuss your project, want to run an idea by him or just want to reach out to a friendly technologist, you can email him at kanwal@2toLead.com. If you don't hear him speak at the many conferences globally, then you can find him active on twitter @kkhipple (over 25,000 followers) or LinkedIn (http://spbuzz.it/linkedinkk).

Chapter 3

Incorporate Metrics in SharePoint Governance to Benchmark, Achieve and Sustain Use

by Jason Schnur

Introduction

Protiviti SharePoint Solutions focuses on helping Clients to achieve and measure value from the investment in SharePoint. Through the course of that work, we have found that the activities necessary to encourage and achieve widespread adoption by Client staff mirror the work done by successful consumer brands to affect choice and influence consumer activity.

It is well known that consumer brands must make significant and continued investments to influence action. Done well, these investments first result in awareness, then in a transaction and, possibly, after sustained delivery of value, brand loyalty. The drive to achieve widespread adoption of Share-Point within organizations requires similar focus and execution. The simple reason? Prospective SharePoint users (all of us) are consumers.

As consumers, we buy for both practical reasons and emotional reasons. The practical ones generally come down to some clear benefit (cost, availability, delivery timeframe, etc.) while the emotional ones are a bit harder to quantify, but very important. For example, when given the choice between Coke and Pepsi, I choose Coke partly because I relate to the polar bear campaign that's been around for almost 100 years. That's a personal choice, of course; some may more closely identify with Pepsi and their "Live for Now" campaign.

The point? Since we are all consumers, we are accustomed to having direct input into the way we spend our limited resources, be they time, money or effort. SharePoint consumers within your organization have that same choice, so the key to unlocking SharePoint's potential and achieving widespread adoption is to approach the work as if the SharePoint intranet were a consumer brand. Let's think in terms of differentiating from competition and helping our consumers to achieve benefit or avoid negative consequences. Most importantly, let's define the specific value that an individual can enjoy by making the choice to actively use SharePoint, as well as the resulting value to the organization.

To do so effectively, it's necessary to document the metrics we want to affect and closely monitor our progress against those metrics as we engage our consumers.

A Consumer Brand to Emulate

Before we consider specific SharePoint metrics in our organization, let's examine the work of a consumer brand that can serve as a representative model. One of the more compelling consumer brands over the past few years has been the automaker Tesla, a California company focused on the idea that electric cars are a better choice than gasoline powered cars. Since beginning operations in 2003 and launching the Model S brand in 2012, the company has grown to over $1.1 billion in revenue and persuaded over 50,000 consumers to purchase an all-electric car (no, there is no hybrid option). Considering that the product can be anywhere from two to four times more expensive than its competitors (the Model S retails for anywhere from $65,000 to $95,000), that is remarkable performance.

Tesla's approach can be instructive to us as we think about convincing our consumers to take action. Specifically, let's focus on Tesla's impact in one geographic market. It's been about 18 months since Tesla introduced the Model S in Norway and, according to the *New York Daily News* in April 2014; it outsold Ford's entire line of cars and sold double the number of Volkswagen Golfs, normally the number one seller in Norway.

To date, Tesla has sold more than 6,000 Model S sedans in a country with about 5 million inhabitants. In fact, during much of 2014, the Model S was the best-selling vehicle in the country, outselling both traditional and electric vehicles.

While Tesla provides an excellent product, there has to be something more to the success it has established in Norway in less than two years. What competitive advantages does Tesla enjoy in that particular market that it has been able to leverage into a leadership position?

First, as reported in November 2014 by *EV News Report*, the Norwegian government has set a goal of 50,000 electric cars on the road by 2017 and is on track to reach that goal sometime in 2015. To support this goal, Norway's government has eliminated its traditional (and very expensive) vehicle sales tax for consumers that purchase electric vehicles. (Considering that, according to The Economist, petroleum accounts for 30% of the Norwegian government's revenue, this is a remarkable initiative. Clearly, the leadership team in Norway sees the value of long-term investment in renewable energy and is willing to take action to realize that value.)

So, consumers can eliminate the tax expense compared to gas-powered vehicles as they consider the value of the Tesla purchase. In addition to avoiding the tax, Norwegians with electric vehicles enjoy other benefits, such as free parking, no toll-road charges and the ability to travel in bus lanes.

Practical cost justification is certainly a way to explain Tesla's success, but there has to be something more than cost considerations. According to an article published in February, 2015 on Teslarati.com—http://www.teslarati.com/norway-loves-tesla-model-s/:

- "Culturally, Norway embraces renewable energy. According to *EV News Report*, 98% of the nation's energy is derived from domestically generated, renewable sources."
- "When asked why he purchased his Model S, an owner from Oslo said, 'The opportunity to drive with a clean conscience.'"

So, in addition to practical cost justification, Norwegian consumers are basing the decision on the emotional reward that comes with using a clean, renewable energy source.

So, in this case an organization (Tesla) took advantage of clear direction by leadership (Norway's government) to convince consumers to take action, both for practical (cost justification) and emotional (environmental effects) reasons.

Let's examine the potential to apply these principles as we focus on achieving and sustaining consumer adoption of SharePoint. To be successful, we need value to be communicated and demonstrated by leadership, we need to focus on the specific value that our fellow consumers can enjoy and we need to appeal to our collective sense of mission, taking advantage of our common desire for accomplishment and the pride that comes with it.

Our Consumer Behavior

We understand that there is a choice in every scenario, and the choice we make needs to be justified based on either achieving benefit or avoiding negative consequence. No one had to teach us this outlook; it's a natural understanding based on our nature and our experience. If we ever touched hot coals, we're unlikely to willingly repeat that same mistake because we want to avoid the consequence. On the other hand, we may be enthusiastic to repeat a decision to volunteer our time based on the benefit it provided to us and others.

We are not too rigid to change behaviors or embrace something new; we just have to be empowered to make an informed decision. And that decision will always be about either achieving benefit or avoiding consequence.

Let's consider this behavior in the context of adopting SharePoint. What clear benefit can be realized or what negative consequence can be avoided? Here are some representative consumer groups within organizations and scenarios that can guide the approach:

- **Project Managers.** They are likely to embrace collaborative SharePoint features if those features can help to achieve project results that are on time, on budget and within scope. If a project team member can't access the project site when outside the corporate network, that one factor may discourage the project manager and team from leveraging SharePoint. If the use of SharePoint threatens to delay a deliverable and damage a relationship, the choice to avoid is easy.

- **Business Developers.** This team may collaborate on proposal documents within SharePoint, but not if that activity takes more time or is more complex than the current process. In that case, it's a clear choice to avoid the potential loss of efficiency that could result from the latest proposal version being "lost" in SharePoint. Negative consequences can be avoided by continuing to use shared drives and email. Securing new clients will always be much more important than attention to document management.

- **Product Engineering Teams.** This group may be spread across multiple continents and time zones. Wouldn't they benefit greatly from having a single, Web-based source for the latest product specifications, design documents and ongoing collaboration among the team? Absolutely, but they will never find out if sites are too difficult to provision or if access is unreliable.

Much like the consumer marketing organizations that can provide a model for us, we'll achieve success by engaging our consumers and determining what is important for them. Like a corporate marketing strategy, the method for achieving sustainable adoption needs to be clearly defined before consumers are engaged.

It's much more difficult to re-engage our consumer colleagues after an ineffective training effort than it is to spend the necessary time up front to make that effort effective.

OK, but we *are* at work

We consumers may collect benefits like an annual salary, a 401K plan, medical expense coverage, etc., so we're not strictly consumers in *all* of our corporate activities. After all, those benefits provided by the employer should assure the organization of something more than fickle consumers whose activity needs to be influenced at every turn by some brilliant campaign, right? (Yes, but only to an extent.) Within enterprise-wide information technology, there are scenarios in which we consumers face no choice other than adoption, for example:

Entering time. Whether this process involves simply "punching the clock" or using an ERP system to track specific activities, corporate consumers use it regardless of the quality of the interface or the ease of use. Why? Because it is simple for us to understand that the activity directly relates to getting paid or invoicing clients.

- The benefit gained? Collecting salary.
- The consequence avoided? Discipline for delaying the invoicing process.

Entering expenses. Anyone who has used personal funds for a business expense is willing to fight through temporary access challenges or an initial lack of familiarity with application features to correctly enter expenses.

- The benefit gained? Recouping funds.
- The consequence avoided? Interest charges, late payments... annoyed spouses.

Tracking Key Performance Indicators. If any part of our earnings are variable and based on achieving specific metrics, we will be happy to document those metrics and log progress against them in a human capital management system.

- The benefit gained? Tracking achievement to receive benefit.
- The consequence avoided? Loss of variable compensation.

We consumers, being generally risk-averse and focused on our day-to-day professional responsibilities, won't embrace the new or additional unless the benefit gained and/or consequence avoided in doing so is made clear. In the above examples, the benefits and consequences are clear.

So, how to communicate and prove to consumers already inundated with choice that SharePoint is worth the time investment?

Our built in advantages

Of course, there is substance to the idea of loyalty to the organization that employs us simply because we are team players. Some of the more self-aware consumers among us may even think "Hey, the organization has been good to me; therefore, I'm willing to embrace this SharePoint thing, even if the value to me right now is fuzzy."

That type of outlook represents a clear opportunity to those of us responsible for influencing SharePoint behavior within organizations, and is a significant advantage that we hold compared with consumer marketing teams. Those teams need to convince consumers that their brand represents better value (more benefit, less consequence) than tens or hundreds of other brands with similar products. But we only need to focus on how our brand (SharePoint) is more effective than a limited amount of known competitors, for example shared drives, email, third-party online file shares, general inertia, etc.

Keep in mind, though, that no one was *hired* to use SharePoint, so goodwill aimed at the company is no guarantee of patience or performance if SharePoint continues to represent a vague value proposition.

Another factor in our favor is that we have a captive audience. There's really no mystery to the size or demographic makeup of this audience. We know exactly *who* we want to influence; we need to determine *what* is beneficial to our audience and communicate those benefits clearly and consistently.

Yet another advantage at our disposal is that the communication opportunity is already built in for us in the form of executive updates, monthly or quarterly meetings, regional roundtables and other common workplace meetings.

Now our path forward to helping our consumer colleagues realize value from SharePoint is becoming clearer. First, understand that we are in a competitive marketplace (for attention), but that we do have some significant competitive advantages, including:

- Ready demographic information.
- Relationships with our "captive audience" consumer colleagues.
- Emotional attachments to our brand held by some of those colleagues.
- Influential leaders who can help to communicate vision and generate awareness and enthusiasm.

Start with Governance

Based on best practices from the consumer marketing world, let's now focus on the methodology you can use to effectively communicate the value of adopting SharePoint by focusing on both practical and emotional benefits. Once you have communicated that value and created awareness, you can move to activities that leverage that awareness and set the stage for long-term adoption.

This methodology is applicable across the SharePoint adoption lifecycle and for organizations of all sizes. For an organization with a handful of staff, these steps can be simple to implement and track. That organization may simply outsource all of the technical considerations by leveraging SharePoint Online within Office365, and focus specifically on the priorities most vital to accomplishing their business goals.

On the other hand, a multinational organization with multiple SharePoint farms requires a more expansive governance strategy that incrementally leads the organization to eventual widespread adoption.

Governance of the SharePoint intranet serves to define, implement, manage and enforce overall site policies. An effective governance plan is important because it provides the "playbook" with which the organization manages priorities and activities on the site. A typical SharePoint governance model includes a cross-departmental team—the Governance Committee—that can represent diverse viewpoints and work together to establish processes that balance enabling consumers and protecting the organization's data and intellectual property.

Main priorities within SharePoint governance plans typically include the following:

- Technical operations
  - Configuration and architecture
  - Environment structure
  - Storage, maintenance and recovery
  - Authentication
  - Support
- Site administration
  - Site provisioning

- Site ownership
- Site content
- Permissions management
- Content administration
  - Overarching information management
  - Content review and scheduling
  - Content organization policies
  - Information architecture modification policies
  - Page management
- Design, personal and social
  - Branding, style and page layouts
  - Personal sites
  - Newsfeeds, blogs and Yammer
- Training processes
  - Site collection owners
  - Site owners
  - Power users

These topics are critical considerations that can help the SharePoint governance committee to manage SharePoint effectively. Taken together, these represent an effective and typical outline. The main problem, though, is that there is little consideration for the majority of consumers within the organization—the business users.

Too many organizations fail to realize the potential value of the SharePoint investment because business users aren't included in SharePoint planning decisions and, therefore, aren't empowered to take advantage of the benefits. According to *AIIM's SharePoint 2013 Industry Watch*, the IT Department is the driving force for SharePoint adoption in 49% of organizations. Only 34% of adoption efforts are business-driven and 14% of those utilize a multi-departmental steering committee.

The graphic below (taken from Protiviti's 2014 IT Priorities Survey) illustrates the significant competing priorities facing the IT team. Yet, despite these competing priorities, nearly half of all organizations still rely on IT to deploy, configure and launch SharePoint, while also training the business user base.

Key Findings from Protiviti's IT Priorities Survey*

Management and Use of Data Assets – Results for CIOs/IT Executives and Large Company Respondents			
Management and Use of Data Assets	Overall	CIOs/IT Executives	Large Company Respondents
Business intelligence and reporting tools	●	●	●
Data analytics platforms and support	●	●	●
Data and information governance program	●	●	●
Data lifecycle management	●	●	●
Master data management	●	●	●
Short- and long-term enterprise information management strategy	●	●	●
Big data initiatives	◉	●	●
End user adoption of data tools	◉	●	●

● Significant Priority
Index of 6.0 or higher

◉ Moderate Priority
Index of 4.5 to 5.9

*Respondents were asked to rate, on a scale of one to 10, the level of importance their organizations assign to each priority. A "10" rating indicates the issue is a high priority while a "1" indicates the issue is a low priority.

It's a daunting proposition for the IT team to assume primary responsibility in these critical areas and oversee the business adoption of SharePoint. It's no surprise that "end user adoption" falls to the bottom of the priority list. In fact, those of us who are directly responsible for achieving Share-Point adoption should be encouraged that "end user adoption" even makes the list at all. Let's add that to our many built-in advantages—an IT team that values end-user adoption and is willing to team with us to achieve it.

Our colleagues in IT are clearly part of our governance committee. Who else should be involved? Most importantly, the committee must be populated by leaders who regularly communicate with our consumers—Communications, Human Resources, and Operations—because effective communication with the greater staff is critical in helping the organization achieve the intended benefits.

Also, the "Training Processes" section of the plan referenced above needs to be expanded to include more than just a focus on training site collection owners and power users. This is an important distinction, given that greater than 90% of all staff in an organization will be SharePoint business users and not site collection owners or even power users. Every step of the methodology that is meant to achieve and sustain adoption, then, needs to take into account the strategy for influencing this great majority of the overall population.

Since the business user community is the largest group of SharePoint consumers, it will have the most impact on the overall success of the solution. Unlike a very specific line-of-business application, the promise of SharePoint is as an *enterprise* platform, so achieving the buy-in of the largest community of users is essential.

Typically, these business users are not engaged as part of the implementation. And if a training plan has been established, it customarily provides information without specific context. In other words, the training is generally heavy on SharePoint features and light on how those features can be applied effectively. Without a consistent user adoption strategy (and feedback mechanism) in place, even the best implementation efforts can be wasted due to lack of adoption.

An effective start to achieving adoption is to include the specific adoption approach directly within the governance plan. So, the Training Processes section above can be replaced with the following topics that include not only the training and adoption focus, but also the intended outcomes of that effort, as gauged by the sample metrics.

- Training and adoption
  - Generating awareness
  - Assessing staff capability
  - Establishing learning objectives
  - Providing training
  - Measuring outcomes against objectives
  - Sustaining adoption
- Sample Metrics *
  - Page views
  - Page views per visit
  - New visitors
  - Average time on site
  - Number of active team sites

Note that these general sample metrics are illustrative and will need to be more specific and targeted to the most important activities at your organization.

Those most directly responsible for adoption within the governance committee then have the responsibility to execute the plan. They do this in concert with IT colleagues on the committee who are focused on other significant initiatives, like protecting the organization's data and intellectual property.

The first step is to plan for the future by committing to:

- Encourage use *and* track results,
- Offer varied learning options (high-impact video, on-demand training, in-person training, active mentorship), and
- Demonstrate flexibility within the plan to accommodate the needs of the entire community (collectively, the consumer base), especially the underserved business users.

Defining the metrics gives us an opportunity to monitor the progress of our efforts and to document that progress objectively for leadership to encourage increased investment in adopting the platform.

Executing the Plan

As organizations consider the effort necessary to achieve and sustain user adoption, they must first specifically define corporate goals for the investment. And then they can encourage the accomplishment of those goals by making the benefits personal to the user community.

Solving the challenge of SharePoint user adoption is not a "one size fits all" approach. Increasing user adoption is a lifecycle made up of practical steps that each business must take to enable long-term success.

Consistent execution of these steps will position the organization well to enable staff and measure the specific Return on Investment (ROI) through establishing metrics and tracking improvement of those metrics. This figure demonstrates the continuous lifecycle of achieving and sustaining high levels of SharePoint user adoption.

The Sustainable User Adoption Model

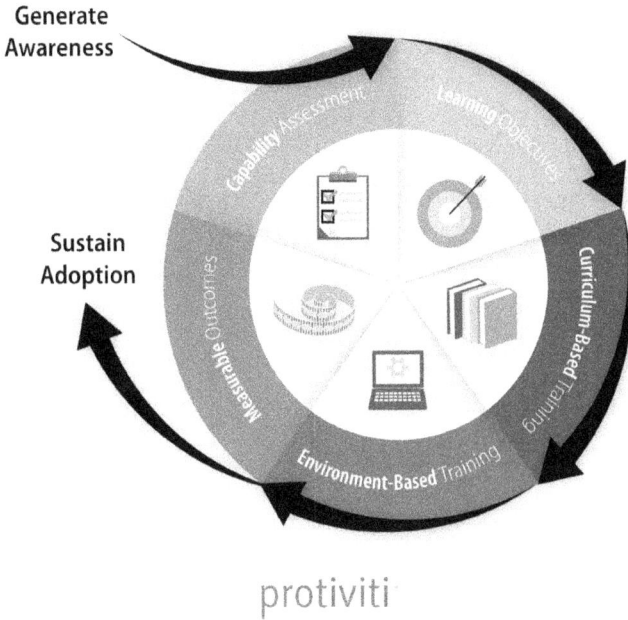

Generate
Awareness

Sustain
Adoption

Capability Assessment

Learning Objectives

Curriculum-Based Training

Environment-Based Training

Measurable Outcomes

protiviti

Generate Awareness by Engaging Staff

The effort involved in generating awareness includes enlisting commitment from an influential leader, gaining agreement within the governance committee on the plan to be executed, and enabling members of your consumer base to demonstrate results (and recruit others).

As in any other team pursuit, leaders have responsibility for defining the vision so that team members are informed and enabled to take action. It is absolutely critical at this stage to engage an influential leader to communicate the vision that SharePoint represents to the consumer community. This leader must be willing to invest the time and energy necessary to both communicate the benefits that effective use of SharePoint can provide and adopt SharePoint to demonstrate its benefits as an example to others.

This same leader should be included as a member of the governance committee and be invested in gaining value from using SharePoint. Choosing the right leader (or group of leaders in larger organizations) is crucial to gaining and keeping the attention of our consumers.

While this is the beginning stage of the adoption lifecycle, the long-term plan needs to be agreed upon by all members of the governance committee. The adoption team has the responsibility for communicating the vision, in line with other governance priorities being concurrently managed by IT and other groups. Once the vision is accepted by the committee, the adoption team can then start to execute the plan.

Let's keep in mind that aiming to achieve sustainable adoption across the entire consumer base in the first attempt is typically unrealistic. We have to plan for changes in strategy based on practical experience, and we have to be flexible enough to make those changes quickly.

The most effective adoption efforts start simply, with a representative focus group that will both gain practical experience by their participation and provide the adoption team an opportunity to improve value to our consumers. That focus group should be chosen because they engage in a high-value activity that can be transitioned to SharePoint (for example, project management).

The focus group will be tasked with adopting (or using more effectively) a SharePoint feature set for the specific purpose of presenting the experience as a case study to their colleagues. When making a buying decision, nothing is more effective to a consumer than a referral from or shared experience of a friend. In addition, it's human nature for us to be invested in a process that involves us.

Enable a team to take action, communicate the results actively and you'll see that others will want to be involved as well. To quote Benjamin Franklin: "Tell me and I forget. Teach me and I remember. Involve me and I learn."

Now let's get the word out to the consumer base and involve them in our plan. This communication should be simple to deliver and simple to understand. The most important consideration is that it should resonate with our consumers, from both a practical and emotional perspective. For example, the practical consideration may involve an effect on clients, as outlined in this message:

SharePoint will allow the organization to more effectively communicate and share information related to our ongoing projects. As a result of adopting SharePoint as our project management solution, we expect to decrease time to market for our products by 15%.

And the emotional message could promote an opportunity to serve the community. For example, Protiviti has launched an initiative for our staff to provide more than one million meals to people in need during 2015. Our collective progress in achieving our goal is updated frequently on the intranet home page, and this initiative has enabled the organization to create a shared sense of community and demonstrate corporate responsibility. Not a bad way to appeal to emotions and instill pride in the consumer base.

Now that we have identified our focus group and the executive sponsor has been able to communicate the plan and the benefits, we can focus on achieving and demonstrating those benefits.

Assess Staff Capability and establish relevant Learning Objectives

Establish a baseline for the initial focus group's capability to use SharePoint. Capability assessments can be completed by creating a simple survey meant to gauge the team's current ability to take advantage relevant SharePoint benefits. If the goal is to leverage SharePoint to manage projects, for example, the questions may focus on familiarity with collaboration tools and document management basics, including metadata tagging, check in and check out, etc.

This assessment sets the stage for reporting progress against objectives and demonstrating success. It also enables us to target specific training elements. In our example, we can provide training to enable the team to effectively use the collaboration features most important for effective project management.

The assessment will also identify specific learning objectives for the training process. The objectives involve filling knowledge gaps across the team so that all are comfortable with the process for effective project management. If the capability assessment identifies a gap in knowledge concerning specific collaboration tools, then one of our learning objectives will address that gap.

Provide training to accomplish the objectives

Training is often the first step that organizations take to help their consumers embrace SharePoint. While this approach seems logical, the training must specifically address an activity that the audience can benefit from, or the first impression we make on that consumer can be negative. Training should be delivered with specific, focused intent. And that intent should be clear to the audience. In other words, help your consumers understand the value of the training, and then deliver that value.

In our example, we have identified the specific processes and features that a project management team should leverage. Before providing the training, we have determined our group's current ability to use those features. The difference between current and desired capability defines our learning objectives and the focus of the training sessions.

Measure Outcomes

It's necessary to understand and document the users' progress in accomplishing the established learning objectives. This is our opportunity to measure the value that has been created as a result of the previous activities and is a practical way to address the metrics we want to influence, which are documented in our governance plan.

One of the main benefits in leveraging a focus group is that the specific outcomes can be demonstrated as part of a case study presentation. These outcomes can be measured simply by re-issuing the original capability assessment to the focus group members and examining the enhanced knowledge that has been gained as a result of the training activities.

This can further take the mystery out of SharePoint and solidify it as a useful resource to help our consumers accomplish their main responsibilities. This case study presentation could be seen by many as the validation needed to start accepting SharePoint as the new way to work.

For many organizations, it may be unrealistic to expect the case study presentation to be viewed live by a majority of staff. But it's simple to record the presentation and make it available on the SharePoint intranet. Now we have an active testimonial that is always available to our fellow consumers.

Continue to Promote and Sustain Adoption

The strategy to this point has been applicable for all organizations. However, critical thought is required to determine the most effective strategy going forward for your specific organization. Will it be best to engage another focus group and continue to build these testimonials? Can you leverage the momentum of that initial credibility and roll out a larger initiative to help a larger percentage of your consumer base?

These decisions, and the pace of adoption efforts, will be unique to your specific organization and should be considered carefully. When in doubt, choose the more modest approach. It's difficult to reverse a failed effort that was widely communicated, and relatively simple to achieve incremental wins along the way.

Summary

Similar to the marketing approach of consumer brands, your approach to SharePoint adoption should be guided by engaging directly with your consumers and tracking the result of that engagement. In addition to working side by side with them, make sure to keep in mind the metrics that were established as part of the governance plan. And track those metrics on the intranet to determine if adoption efforts are having the intended effect.

Our illustrative metrics were:

- Page views
- Page views per visit
- New visitors
- Average time on a site
- Number of active team sites

What is the baseline you established for each? What improvements to that baseline will represent success? In what timeframe should these improvements be realized? Specifically, what activities that can be tracked will represent success?

And, finally, what other metrics should be added that will provide even more insight into the value that SharePoint is enabling at *your* organization?

The ability to gain those insights will be determined by the effectiveness of your upfront planning. Start with documenting the metrics within the governance plan. Then develop your consumer marketing strategy and implement it consistently. And, if you become discouraged along the way, relate your situation to consumer brands like Tesla that have faced challenging circumstances and have risen to become market leaders.

Like these innovative brand pioneers, you can succeed through understanding the value that consumers seek (both practically and emotionally) and delivering that value with passion and purpose.

ABOUT THE AUTHOR

Jason Schnur leads the Governance, Training and Adoption practice within Protiviti SharePoint Solutions. His passion lies in helping Clients to achieve and sustain adoption of the Microsoft SharePoint platform across their entire user community.

He has nearly 20 years of experience focused on enterprise technology, including CRM, ERP and enterprise content management applications, and regularly shares his experiences on Twitter (@jasonschnur) and on Protiviti's SharePoint blog: http://sharepoint.protiviti.com/blog.

Jason earned an international business degree from Penn State University and lives in Falls Church, VA with his wife and two children.

Chapter 4

Real World Activities:
How Organizations Drive Adoption

by Richard Harbridge

Change isn't easy. We know that with the right planning and activities we can accelerate and guide change. With the right effort, we can increase adoption of Office 365 as well as help foster the new behaviors that technology enables. In this chapter, we will explore why change is challenging and provide a significant number of activities you and your organization can leverage that will help drive adoption.

Change Isn't Easy. It Needs Our Help.

Change isn't easy and it doesn't come naturally. Even when there is a good reason for it. Even when change can mean life or death. Let me share a short example from a research study, which can be found in a great book called "Immunity to Change." [1]

Doctors told heart patients that they will absolutely die in 6 months if they don't change their habits around lifestyle, diet, and exercise. Those individuals certainly had the motivation and a sense of urgency. They even had the tools and the vision as well as an understanding of the end state. Yet in this situation, only 1 in 7 or 14% could manage to carry out the change on their own. Desire and motivation aren't enough on their own: even when it's literally a matter of life or death, the ability to change remains maddeningly elusive.

Think about that. If one in seven people with an absolute need to change can't change, what chance do we have that individuals in our organizations will change how they collaborate or communicate on their own? What are the odds of them adopting Office 365 and new ways of working without help? The answer? Very low odds indeed.

The good news is that there are activities we can perform that provide the incentives and help that your users need to embrace change. In this chapter we will illustrate many of these activities and initiatives that we can implement to help drive change in our organization and in the behavior of our users. The bad news is that change is hard and it won't happen without effort from leaders like yourself.

So How Do We Help Encourage Change?

How we help encourage change is dependent on many factors, such as the kind of change we are driving and situational factors like the environment, the individual, and circumstances surrounding the change. The goal of this chapter is not to provide generalized guidance on driving change (there is already plenty of great material out there on this topic), but to give action-able examples and activities that other organizations have tried that have driven or improved adoption. In other words, "Don't tell me that change management and adoption is important. Tell me *how* we can improve adoption, and provide specific examples." Well, that's what we are going to review.

Real World Adoption Activities & Examples

Years ago when I worked within a large enterprise, we knew that building great solutions leveraging SharePoint would have a very positive impact on the business. After building great solutions, we quickly realized that many users weren't even using the basic functionality of the platform, and we had to rethink our focus. In order for us to realize the value we wanted, we had to get our users to adopt the new technology itself and change the way they worked.

What we learned was that some adoption activities had more impact than others and that half of the challenge was answering the question "what can we do together to help encourage change?" Years later, I have had the luxury of consulting with hundreds of organizations to build adoption roadmaps and strategies. What we find is that defining and executing the right activities makes all the difference. I am sharing the most successful adoption activities in this chapter. They were discovered based on actual experience, extensive research, assessments and work done with Microsoft and analyst agencies.

These activities are based on real-world successes (though they vary in impact). The activities may help your organization be more successful in driving adoption of Office 365 and related technology like SharePoint Online, Lync Online and Exchange Online. The activities also translate to on-premises server technology like SharePoint 2013, SharePoint 2016 and more. At the time of this writing, there are document templates and resources at http://Success.Office.com that Microsoft has prepared to help you accomplish these core Office 365 adoption activities as well as many of the other adoption activities listed in each activity section.

Author's Note: *Keep in mind that this is a subset of adoption activities Microsoft has seen in practice and that there are always new and creative ways that organizations find to drive adoption as the technology and industry change. If you are looking for more beyond this list, be sure to check Microsoft resources and community resources. Also feel free to reach out to me on my website http://www.RHarbridge.com. I will actively work to share additional templates, examples and activities based on demand from wonderful people like you.*

Pre-Work Recommendations

There are certain activities that can significantly improve your own organizational understanding and lead to much more effective adoption planning. Some organizations may be more prepared for an effective rollout of new technology than others. My aim in this section is to explain a few of the more critical or overlooked areas of pre-work that you may want to invest additional time or effort into.

Workforce & Persona Analysis

Workforce analysis is a fundamental part of understanding the impact you can drive with your users and within your business. It helps inform how your users work across regions, in business units, or in different roles. We want to dig into collaboration, innovation, communication, device preferences, location and other areas that are most impacted by Office 365.

One of the reasons this is considered a pre-work recommendation is that building personas greatly helps in the adoption planning process and benefits from workforce analysis. Personas help facilitate a shared understanding of your target user base and provide a means to better communicate needs and characteristics that should be considered when planning for adoption.

Lastly, when you combine the personas with scenarios you have identified that should be prioritized, you will be able to more effectively plan your adoption activities and maximize your impact.

Current State Analysis

It is extremely important to have a set of baselines of:

- How people work today
- What systems they use
- What challenges currently exist

Current productivity levels or the time it takes to execute key scenarios that will be impacted by the adoption of Office 365 (optional but recommended)

Keep in mind that one of the activities you will execute is defining your success metrics based on goals and scenarios you map to Office 365 solutions. So this is just meant to compliment and inform the activities that follow.

Rewards & Recognition Plans

How are employees recognized today? Are there ways in which you already recognize individuals sharing, collaborating or supporting one another? To effectively leverage a move to empowered modern communication, collaboration and productivity technology, your organization will need to facilitate a culture change that supports the new ways in which people can work. I am calling out the rewards and recognition plans that exist today in organizations as being the most impactful (and representative) of the company culture. These can also be a great way to tie adoption to broader company vision and culture enablement.

Adoption Roadmaps
I list many adoption activities in this chapter. Create your own practical roadmap to define the order in which they are executed, their dependencies and the critical path for success. This step helps ensure there is strong coordination, shared understanding and shared commitment to achieving the targeted adoption goals you set out. If you have created a roadmap before, it will certainly help and can be used as a base to build out your adoption plan for Office 365.

Understanding the Capabilities of Office 365
To effectively drive adoption of any technology, it is important to first understand what it is capable of, how others are using it successfully, and when it makes sense to use one capability over another.

If you don't have this expertise, it is highly recommended that you work with an experienced expert who can help educate and guide you as your prepare your adoption plans.

Phases of the Rollout

It's also important to understand the phases that are a part of any new technology rollout. Within each activity description, I have identified the phase in which it is typically executed. These are just recommendations. In many cases, you may repeat an activity, or adjust the timing of it based on your own circumstances.

Pre-Launch: The pre-launch phase is where you identify what is needed for a successful launch and create your plan. During the pre-launch phase, you prepare for the launch and ensure that effective testing, pilot or feedback opportunities exist.

Launch: This is the actual launch initiative. During this phase it is important to capitalize on pre-launch work and take full advantage of the launch itself to kick-start and begin driving adoption.

Post-Launch & Ongoing: This is without a doubt the most commonly neglected phase. It is critical to invest time and energy after the launch of Office 365 to ensure it is meeting user needs, helping users, and that users are effectively adopting the technology. So much energy is put into getting to the launch that it's easy to forget about the work that comes after it. Be sure to carefully plan and execute post-launch and ongoing adoption activities for as long as the technology is being used! Even if you have good adoption rates, these activities can be important in helping to maintain adoption.

As you read through the activities that follow:

- Consider the relevant activities for your leadership group.
- Consider the activities that drive awareness and usage.
- Ensure that you have a plan for how to effectively help users understand the value that the new technology provides.

Each activity has been listed in the category that maps closely to one of those primary purposes.

Leadership Activities: Making Office 365 Successful

Before we explore creative or interesting ways to drive adoption, we must start with fundamental steps and activities that should always be accomplished first. What follows are examples of activities that are highly recommended for every organization to complete before investing in additional adoption activities.

Identify Key Stakeholders	
Suggested Timing: Pre-Launch	**Level of Effort:** Low

The most successful Office 365 rollouts occur when a committed team of individuals, representing a cross-section of the organization, execute all tasks effectively and on time. You need to identify stakeholders and ensure that they have clear expectations of their time commitment, qualifications and responsibilities.

Examples of key stakeholders:

- *Business owners*
- *Champions*
- *Communication leads*
- *Community managers*
- *Department leads*
- *Executive sponsors*
- *HR managers*
- *IT specialists*
- *Project managers*
- *Training leads*

Engage Executive Sponsors

Suggested Timing: Pre-Launch **Level of Effort:** Low

It's essential to obtain the buy-in and support of leadership prior to introducing Office 365.

During the preliminary phase, executive sponsors should:

- Help the project team(s) craft the overarching vision for Office 365 by tying it to broader organizational objectives.
- Play a role in communicating the vision to other leaders across the organization as well as employees within your organization.
- Actively and visibly participate and use Office 365 capabilities to reinforce desired behaviors and help drive adoption by end users.

Create A Well-Defined Vision for Office 365

Suggested Timing: Pre-Launch **Level of Effort:** Low

A well-defined vision enables employees of all levels to foresee the value the new tools will bring, not only to the organization as a whole, but also to the individual roles within it. This in turn helps secure buy-in and support across the business.

Here are two simple examples of a vision statement:

To transform the way people within our organization connect with each other, so that when we work together on documents, tasks or projects we can improve collaboration and tracking and reduce redundancy and extra steps in relevant processes.

To transform the way our department:

- Connects with each other,
- Works together on tasks or projects,
- Communicates with leadership, and
- Streamlines processes for efficiency and clarity.

Determine Actionable Goals & Map to Solutions in Office 365

Suggested Timing: Pre-Launch & Ongoing **Level of Effort:** Medium to High

You'll want to understand the common business challenges experienced across departments and teams, which in turn will help you identify the practical solutions.

Host a meeting with your key department stakeholders, project managers and business champions to define the business processes that will help them meet their broader business goals and prioritize the Office 365 solutions that will help them achieve those goals.

Define Your Metrics for Success

Suggested Timing: Pre-Launch & Ongoing

Level of Effort: Medium to High

When you develop your ideal scenarios and solutions, it's critical to come up with a formal set of success criteria to measure the impact resulting from your Office 365 rollout. You'll need to determine what should be measured and how you will go about collecting both quantitative and qualitative data.

Choose criteria that will help you showcase success to leadership, such as user satisfaction, employee engagement, adoption velocity and figures related to your desired business scenarios.

Create Your Communication Plan

Suggested Timing: Pre-Launch & Ongoing

Level of Effort: Low

Communication is important throughout the process of launching new workloads or Office 365. Plan major milestones and critical communications so that all stakeholders understand the communication plan.

Train Your Helpdesk

Suggested Timing: Pre-Launch & Ongoing

Level of Effort: Low to Medium

As you begin launch planning, a fundamental necessity is ensuring that your helpdesk is ready to support basic Office 365 scenarios and that you are ready to manage provisioning of licenses, rights and other key essentials.

Awareness Activities: Helping People Use Office 365

These activities are designed to get people to adopt Office 365 and help them get started using the technology. Because end-user education is critical to success, this section provides some training and education-based activities. Included with each activity is a short description of what the activity is and how to implement it, along with some real-world examples of how it works in practice. *Keep in mind that the goal is to drive adoption and not just usage (adoption goes beyond usage).*

Leverage Champions

Suggested Timing: Pre-Launch & Ongoing   **Level of Effort:** Low

Pre-Launch: Identify "champions"—the top active contributors and users of the new technology. The best champions are those that are generally optimistic, like to share, and are very responsive to their peers. Support, train and engaged these individuals to effectively influence organizational readiness, adoption and experience with the new technology.

Ongoing: Keep in mind that you have access to all sorts of usage and signal data in Office 365 that can help you determine who active users are and which users have considerable influence and impact. These individuals often require or can benefit from additional support, guidance, training or engagement from your team.

Create Countdown Email Campaigns

Suggested Timing: Pre-Launch   **Level of Effort:** Low

Send a "Countdown Email" to let your audience know what's coming, set expectations and spark interest by focusing on the "What's in it for me?"

Broadcast Portal Announcements

Suggested Timing: Pre-Launch & Ongoing   **Level of Effort:** Low

Work with Internal Communication to make announcements across your company portal and/or IT portal as needed.

Post Internal Advertisements

Suggested Timing: Pre-Launch   **Level of Effort:** Low

Ensure that Office 365 has a visual presence throughout your corporate offices with posters, flyers, educational booklets and other print messaging.

Portal ads, posters and more advertising assets like t-shirts can be found at the Microsoft Office store: http://www.co-store.com/iw.

Show Teaser Videos

Suggested Timing: Pre-Launch   **Level of Effort:** Low to Medium

Play teaser videos during pre-launch events, perhaps by the elevators or by the cafeteria. It's a great way to generate buzz and excitement. Ellen van Aken has an amazing collection of intranet launch videos you can use for inspiration here: http://www.scoop.it/t/intranet-launch-videos-and-teasers

Stage an Awareness Event

Suggested Timing: Pre-Launch   **Level of Effort:** Low

Host an in-person event where users can discover Office 365, talk to a project team member at various scenario stations, and access training resources. Have the event in a high-traffic area such as a lobby or lunch room.

Send Announcement Emails

Suggested Timing: Launch   **Level of Effort:** Low

Send out an "Announcement Email" to let users know what's available, how to get started, and where to go to find help and resources.

Host a Launch Event

Suggested Timing: Launch   **Level of Effort:** Low to High

Host a large-scale launch event, such as a company all-hands or town-hall-style meeting, in which the executive sponsor and rollout team can officially introduce Office 365 and discuss the value proposition.

Conduct a Baseline Survey & Follow-Up

Suggested Timing: Launch   **Level of Effort:** Medium

This activity is complementary to announcement emails.

Circulate a baseline survey shortly before users receive activated accounts and devices, to gather data about their knowledge of Office 365. Based on the responses to this survey, provide targeted follow-up recommendations and next steps. Highlighting resources they can access, such as the training site, upcoming webinars and events, and useful "getting started" resources.

Send Weekly or Biweekly Tips & Tricks Emails

Suggested Timing: Post-Launch & Ongoing   **Level of Effort:** Low

Periodically share tips with end-users by using "tips & tricks" emails to sustain momentum and broaden the use of each applicable Office 365 scenario.

These don't have to be limited to emails either. If you are implementing a portal, why not put the tips and tricks on the portal? Or on Yammer? That way, the tips are reusable and searchable.

Include an FAQ Listing & FAQ Finder

Suggested Timing: Pre-Launch & Ongoing   **Level of Effort:** Medium

Often it can be daunting and troublesome learning a new system, process or technology. By adding a simple and easy-to-read Frequently Asked Questions (FAQ) section to your portal, you can remove some of the confusion and help reduce the amount of simple reactive question support you provide (to focus this energy on more proactive and useful tasks).

What's more, if you implement content types and consistently enable or encourage the creation of FAQs across various systems, processes, policies and teams in your organization, you can create FAQ collections or FAQ finders.

As a minimum, create and maintain an FAQ list to address the most anticipated questions. Post the FAQ on your internal site, or Yammer group, and assign a team to update it regularly.

Many organizations leverage SharePoint lists to enhance or provide improved FAQs. Leveraging a SharePoint list enables you to track the last modified date of each individual FAQ (question and answer). You can also see who last modified that question/answer, categorize them, and provide more advanced visualizations of the FAQs on targeted sites by rolling up subsets of the FAQs.

Schedule Buzz Days (Continual Awareness Events)

Suggested Timing: Post-Launch   **Level of Effort:** Low to Medium
& Ongoing

Periodically host in-person events (biweekly or monthly Buzz Days) where users can browse Office 365, talk to a project team member at various scenario stations, and access training resources. Have the event in a high-traffic area such as a lobby or lunch room.

Conduct User Experience Surveys

Suggested Timing: Post-Launch & Ongoing   **Level of Effort:** Low

You can release a survey halfway through your launch to gather data about users' experiences with Office 365 and use the results to make any necessary adjustments.

After your organization-wide rollout, conduct a final launch survey to assess user satisfaction. You can release this survey 90 days after launch, and then in quarterly or twice-a-year intervals continue to measure user adoption from a satisfaction and productivity standpoint.

Showcase Spotlight Days

Suggested Timing: Post-Launch & Ongoing   **Level of Effort:** Low to Medium

Capture success stories and showcase them through "Spotlight Days," where an employee or team is recognized for their successful use of Office 365.

Report on Success Metrics

Suggested Timing: Post-Launch & Ongoing   **Level of Effort:** Low to Medium

After launch, it is surprising how many organizations don't do a good job of reporting on improvements, changes, and progress toward achievement of the initially-identified success metrics.

Track results based on your previously defined success metrics, and measure progress against your benchmark. Periodically report results to key stakeholders.

Authors Note: Make sure that one of your early surveys enables you to capture a baseline for the business problems that you are trying to solve. These follow-up surveys are about the business problems, and you need to have a baseline to measure impact. The key issue is not simply adoption per se, but how adoption of the technology drives business results.

Create a Feedback & Improvement Yammer Group

Suggested Timing: Launch & Ongoing   **Level of Effort:** Low

Make sure to encourage your users and champions to develop ideas for how Office 365 can improve business practices and to share them with others via a Yammer group. Use these ideas to generate additional usage scenarios and to kick off additional training initiatives.

Incorporate Daily Performance Metrics into Your Portal

Suggested Timing: Post-Launch & Ongoing   **Level of Effort:** Low

Providing daily performance measures—especially for retail, sales or help desks—can be a great way to get people to access the intranet regularly. This performance information tells your users exactly how they are doing against personal or organizational performance goals and will drive those users to the intranet or portal.

Incorporate Engaging Content into Your Portal

Suggested Timing: Post-Launch & Ongoing   **Level of Effort:** Low

Believe it or not, sometimes the most important thing is getting people to regularly visit a SharePoint site or spend time in the Office 365 experience, to become more familiar with it.

Finding frequently used (daily) content can help make leveraging Office 365, or certain components of Office 365, more common and habitual, and ease people into using it regularly.

Here are some examples of incorporating engaging content into your portal:

1. *Cafeteria & Lunch Menus:* One of the most popular items in a portal or intranet on Office 365 is the cafeteria menus or menus from favorite lunch destinations. In one organization, they even provided a "delivery to your desk" service through their portal where staff could order food from the cafeteria to be delivered to their desk.

2. *People on the Move:* You can provide another reason for people going to the intranet by posting announcements of promotions, new hires and people who have found a new role in the organization.

3. *Employee Milestones:* Anniversaries and years of employment are something every organization should have in their recognition and reward programs. Integrate or showcase those in your Office 365 intranet.

4. *Personal Milestones:* To make the portal or Yammer more engaging and personal, you can encourage employees to share significant personal events and milestones, such as the birth of a child, an engagement or a marriage.

5. *Inspiring Quotes:* To make the portal feel more engaging, you can cycle famous quotes. Keep content fresh and inspiring. This can add to motivation and provide one more reason for employees to visit the intranet portal.

   A similar activity some organizations incorporate is to add a humor element by sharing a common "filler word" each week. These filler words or phrases are commonly used expressions in the organization, and they can be featured in a short sentence or story. This can be a way of sharing a company inside joke, or acknowledging or playfully poking fun at an aspect of your company culture.

6. *Weather, World Clocks, and Stock Tickers:* This one pops up often as a requested item on portals to help share at-a-glance views of certain information. While this won't change behavior, it can be one more reason for navigating to an Office 365 portal environment.

   One of the common difficulties users have when interacting in a global company is determining what time it is in other offices. Consider incorporating a third-party component or free technology to create world clocks on your portal. Rather than having to search Lync contact cards or ask which time zone Office XYZ is in, employees will have one more reason to visit the intranet.

7. *Event Photo Sharing:* After a significant organizational event, it can be useful to share photos of the event. This can lead to people (typically those who attended the event, or who know other employees who attended) reviewing and interacting with the photo content. You may want to create tagged repositories of photos, depending on your internal policies and culture.

8. *Employee Discounts & Offers:* Many organizations have discounts through the relationships they have with various vendors and institutions. Highlighting and sharing these in an accessible way on the intranet portal can bolster adoption.

Create Mock Profiles

Suggested Timing: Post-Launch & Ongoing   **Level of Effort:** Low

This is an unusual activity that worked for a company that wanted to incorporate humor and personality into their launch activities and Office 365 portal. In this organization, they created a mock profile of a celebrity user and uploaded documents and pictures to add to the profile. The profile served to create humor, to comment on things throughout the organization, and also to drive users to understand how much an individual can do in Office 365. You could try simple models of this with a fictional character profile that represents a product, your portal brand, or a seasonal event—for example, a Santa Claus "profile."

Create Real Organizational Profiles

Suggested Timing: Pre-Launch,   **Level of Effort:** Low
Post-Launch & Ongoing

While mock profiles are interesting for getting people engaged, there is often a very real need for authoritative profiles that represent key business interests.

In most organizations, the content uploaded by a head of a department or group may be fantastic. But if that person changes roles or if their focus shifts, the content that they originally maintained, updated, commented on, or contributed to may also change.

For continuity and to show clear authority, some organizations have created user profiles that represent key offices or personas. Examples include "The Office Of Contoso" or "Contoso HR"—which might share key announcements, official documents, or official answers/responses to posts on Yammer. You can use these profiles to officially curate, publish, engage or share information.

These real organizational profiles can help drive adoption by providing additional ways for a user to engage the organizational entities directly. The profiles also provide frequent updates to content and can provide new ways for offices, departments or teams to engage users.

Incorporate Profile Links in Employee Signatures

Suggested Timing: Post-Launch & Ongoing   **Level of Effort:** Low

If key people add their internal profile link to their Delve profile or Yammer profile, or incorporate a link to an intranet site they manage into their email signature (when emailing internal people), this can help encourage people to visit and view each other's profiles or shared content.

Hold Contests & Competitions

Suggested Timing: Launch, Post-Launch & Ongoing   **Level of Effort:** Low

Planned contests can help to drive interest and awareness. Scavenger hunts, trivia contests and even competitions can be a good way of engaging people. But they often don't drive sustained adoption, so it's important to keep these as awareness activities.

One organization ran a SharePoint Olympics event, and another ran a Productivity Olympics competition that included multiple events and leaderboards over a period of weeks.

Run "For Charity" Reward Campaigns

Suggested Timing: Pre-Launch, Launch,   **Level of Effort:** Low
Post-Launch & Ongoing

Consider rewarding an activity that you want people to do on the portal by making a contribution to a charity.

Example: One organization ran a competition during their United Way campaign to encourage users to update their profiles in SharePoint. The department with the highest rate of user profile updates earned bragging rights and the ability to select a targeted charity that would get a $1,000 donation.

Define Portal & Site Identities

Suggested Timing: Post-Launch & Ongoing   **Level of Effort:** Low

Giving key SharePoint or portal sites an identity, or using vanity URLs, can help people refer to them by name and help spread awareness. In most organizations, the intranet has a branded name. Similarly, some organizations have large sites or other portals that they name to help promote recognition of, and pride in, the site.

Incorporate Feedback in Portal

Suggested Timing: Pre-Launch, Launch,   **Level of Effort:** Low
Post-Launch & Ongoing

As you release new features and functionality, provide multiple ways for users to provide feedback and for your team to take action and effectively address the feedback.

Consider not just incorporating feedback on a new portal, but in your enterprise search results pages as well. So if users cannot find content, they can provide feedback that will enable you to improve the search experience.

Example: One organization provided a service-level agreement (SLA) on feedback for search results. When users commented that they couldn't find something, IT ensured it displayed in search results within a certain amount of time and provided a notification to the requestor that the item now shows up when they search (with a link to the search query and page).

Create Baited Email Hooks

Suggested Timing: Post-Launch & Ongoing **Level of Effort:** Low

We want to ensure effective communication; but sometimes it's better for someone to read content in a curated portal experience or in an updated document than in a short email. In these situations, you can purposefully and strategically omit certain information from an email announcement to drive the reader to visit the site. This can serve the dual purpose of getting employees to visit Office 365 and ensuring the content they receive is the latest and most complete.

Remove Alternatives & Make It Mandatory

Suggested Timing: Launch, Post-Launch & Ongoing **Level of Effort:** Low

This stance can make many companies uncomfortable, but it can be a sure way to encourage use of the platform instead of legacy alternatives that are less secure or less integrated.

This should be used in combination with guidance and education materials that help users understand how to use the technology effectively.

Education Activities: Helping People Understand the Value of Office 365

These activities are designed to help people understand the value that an Office 365 feature or capability can bring to their daily work. These activities can also be leveraged to help people optimize their usage of Office 365.

Create Pilot & Early Adopter Yammer Groups

Suggested Timing: Pre-Launch **Level of Effort:** Low

Create a Yammer group to engage with pilot members and anyone interested before launch.

Create Training Site

Suggested Timing: Pre-Launch **Level of Effort:** Low

Use an internal team site to store training resources, such as getting-started guides and tips & tricks. You can also direct users to the Microsoft Office 365 Learning Center: https://support.office.com/en-US/learn/office365-for-business.

Connect, Teach & Empower Champions

Suggested Timing: Pre-Launch, Post-Launch & Ongoing **Level of Effort:** Low to High

Learning with and from coworkers is one of the most highly effective and recommended methods your organization can leverage.

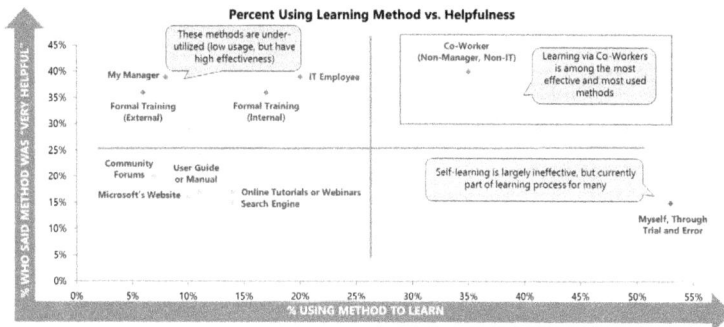

To maximize the impact, frequency and effectiveness of this form of learning, create a community of "champions" or early adopters who can be engaged to provide field-based support across the organization. Provide advanced training and an advanced escalation path for issues identified by your champions, so that they can promote the solution to their peers and colleagues. This ensures that champions remain supportive and effective at teaching others in the organization by example and by direct coworker interaction.

Hold In-Person Training Events

Suggested Timing: Pre-Launch, Post-Launch & Ongoing   **Level of Effort:** Low to High

Host an in-person event to train champions and pilot members. Use customizable training decks mapped to the most fundamental Office 365 usage scenarios.

Conduct Baseline Survey on Office 365 Understanding

Suggested Timing: Pre-Launch   **Level of Effort:** Low

Shortly before champions participating in your soft launch receive activated accounts and devices, circulate a baseline survey to gather data about their knowledge of Office 365.

Announce & Explain New Features & Capabilities (Office 365 Service Update Announcements)

Suggested Timing: Post-Launch & Ongoing   **Level of Effort:** Low to Medium

Keep an eye on Office 365 service updates and inform people of new features and updates as they are released.

Providing regular improvements means there is always something new to try or new uses to share that can spark interest. Announcing major and minor releases of functionality can add a sense of continued investment in the platform from the business and provides new opportunities to engage users in better ways they may be able to work.

Be sure to leverage Microsoft new update releases in this way. They can be found at *http://Roadmap.Office.com.*

Conduct Champion Experiences Survey

Suggested Timing: Pre-Launch   **Level of Effort:** Low

Optimally, you would perform multiple surveys beyond the baseline to help guide improvements to awareness and training materials.

Release a survey halfway through your soft launch to gather data about champions' experiences with Office 365, and use the results to make any adjustments prior to a general rollout.

Use a final survey immediately after the champions' soft-launch period to determine whether you need to make further adjustments to your general training and awareness materials.

Communicate Policies & Guidelines

Suggested Timing: Pre-Launch, Post-Launch & Ongoing   **Level of Effort:** Low to High

Along with end-user training, be sure to communicate your organization's specific policies and best practices, so users are aware of specific guidelines and how they're expected to use Office 365.

Share Summaries of Top Searches & Corrected Failed/Abandoned Searches

Suggested Timing: Post-Launch & Ongoing   **Level of Effort:** Low to Medium

For many organizations, the launch of Office 365 or a new version of SharePoint is a great opportunity to improve and encourage the use of search capabilities. To that end, it can be very useful to share the top searches with users so that they know that search is being leveraged, and can use those same popular searches to find content that can help them.

It can be useful to have a search team. Many organizations assume search just works, but it often requires monitoring and improvement over time by individuals within the organization.

Share that the search team is listening to feedback and watching for searches that return 0 results or no useful results (abandoned searches). Correcting the search so it returns the right results, and surfacing the corrections, can encourage many to reinvest time and energy into the search experience.

Review Usage Reports & Relevant Audited Activity

Suggested Timing: Post-Launch & Ongoing   **Level of Effort:** Low to Medium

Measure against the success metrics you put into place by leveraging the existing reporting found in SharePoint and Office 365. Additional third-party technologies can provide advanced reporting beyond what the Microsoft technology offers out of the box. These auxiliary technologies may be worth considering to compliment your efforts in better understanding activity, adoption and engagement of Office 365 and SharePoint.

For example, it may be useful to track usage of key sites to determine the number of times certain content is shared within a specific location or to summarize more broad results such as the growth of documents being stored across OneDrive for business locations.

These reports can also help improve targeting for champion discovery, as very active users may be good targets for additional training, support or engagement.

Incorporate Portal into New-Hire Orientation

Suggested Timing: Post-Launch & Ongoing   **Level of Effort:** Low to Medium

Incorporating new-hire orientation into the portal can be a great way to introduce the portal and make the onboarding process more engaging. The system can help them fill out their user profile, interact with content in various sites, and understand what technology to use when.

Incorporate Portal into Annual Training or Skills Validation

Suggested Timing: Post-Launch & Ongoing   **Level of Effort:** Low to Medium

Some organizations require completion of in-person or online classes by a set date. By adding Office 365 training into the annual training and skills validation process, you are showing leadership support and authority on the importance of the new technology or system.

Collect, Share & Support Success Stories

Suggested Timing: Post-Launch & Ongoing   **Level of Effort:** Low to High

While many may not see success stories or qualitative feedback as training material, often with minor effort a real user story can be made into a training scenario that is not only easily understood but is an exemplary way of working and realizing impact from the use of the technology involved.

Adding a button or option to enable users to submit success stories can lead to many people initiating a submission or sharing success messages. These stories not only show the value of the technology, but help provide personalized insight into how impactful it can be.

In addition to giving people the option to share stories, make sure to present these stories effectively to users in your organization and to spend energy discovering new ones.

In one organization, each year before they went into their IT strategy and vision meetings, the team spent a week or two leading up to the meetings collecting feedback, stories and relevant information from their users. This empowered them to celebrate and showcase technology-aided successes in the past year, and also helped identify user needs, challenges and opportunities for IT to invest in for the coming year.

Hold YamJams, Use #YamWin & Create Starter Yammer Groups

Suggested Timing: Pre-Launch, Launch, Post-Launch & Ongoing   **Level of Effort:** Low to Medium

There are a few Yammer groups that most organizations implement before the launch of Office 365 and a new wave of collaboration and communication technologies that include Yammer.

Create a Yammer Help Group that can act as a place for people to share their questions and key resources, and to create a helpful Yammer community that scales your IT organization's ability to support users. This can also be a great place to discover, identify and recognize champions.

As your coworkers use Yammer, ask them to think about ways that Yammer has helped them. Then encourage them to describe their experience and post their stories to the company network with the hashtag #YamWin.

Afterward, find the best YamWins and post them to the Yammer Help Group as examples to inspire your entire company.

Consider creating another group, called the Yammer Ideas Group. Use this group to brainstorm ways that your company can use Yammer to improve how work gets done. Ask your coworkers to share their ideas in the group. Then test out the ideas to see how they work. Showcase examples of success. Brainstorm ways to improve ideas that need more work.

Consider holding YamJams, or at least an initial few, as you launch the new platform in your organization. Engage one or more executives or department heads to participate in 30–60-minute, live discussions using Yammer. Discussion topics will be predefined and a Yammer community manager will help facilitate and moderate.

Provide Best-in-Class & Solution Showcases

Suggested Timing: Launch, Post-Launch, & Ongoing

Level of Effort: Medium to High

Often one of the best ways for a user to learn is by example. So as individuals or teams of people in your organization build effective team sites, user profiles, groups, solutions, or more, be sure to recognize them, showcase them and share why they are such great examples that others can emulate or use to improve.

This is also a great way of recognizing the impact that your champions are driving throughout the organization in helping people connect further.

Example: One organization started with this simple idea and added an official logo image to each site or solution that was a best-in-class example. Over time, they created a directory of these best-in-class examples. The solutions found within this directory became so popular that they expanded the directory into a miniature marketplace where business units could engage with one another and share how they built a solution. Some units would even provide services or support to enable another team or group to leverage their solution. IT expanded on popular solutions so that they could scale more effectively and better ensure that all existing solutions received updates and improvements in a more centralized way.

When IT or the business develops new solutions that leverage SharePoint or Office 365, showcasing the solution can be extremely useful. Highlight who owns it, what its purpose is, what benefit it has provided (ROI would be great), screen shots, and more. This often leads to two useful things:

- Users engage with one another, which decreases cost of ownership for IT and expedites the requirements gathering process (when the business user says "I want that").
- It enables and encourages those who have adopted or are using the platform to share their successes, be proud of them and present them in a consistent way.

Provide Community Showcases

Suggested Timing: Post-Launch & Ongoing

Level of Effort: Low

As your organization matures and begins to engage with external communities or builds and fost ers its own internal communities, promote these communities to other employees. This can lead to more adoption of both the community and the portal as a hub for many of these formal and informal community activities.

Provide Guidance on What Tools to Use for What Purpose

Suggested Timing: Launch, Post-Launch & Ongoing **Level of Effort:** Medium to High

The goal of providing guidance on what users can do is to highlight the many things they can do, as well as to demonstrate to them how and when they should use the technology. This graphic presents a very basic starter list of activities a user can accomplish with Office 365.

Consider a relatively simple scenario, such as collaborating on techniques outlined in this chapter by writing them together and discussing them. Should we perform this activity in One-Drive for Business, in Outlook via email attachments, via a Lync meeting, in Yammer, or in a SharePoint team site specifically for this purpose? In this scenario you have a plethora of ways to collaborate, each with benefits, and each with different experiences. In many cases you might perform this activity in multiple ways that combine together into a better result.

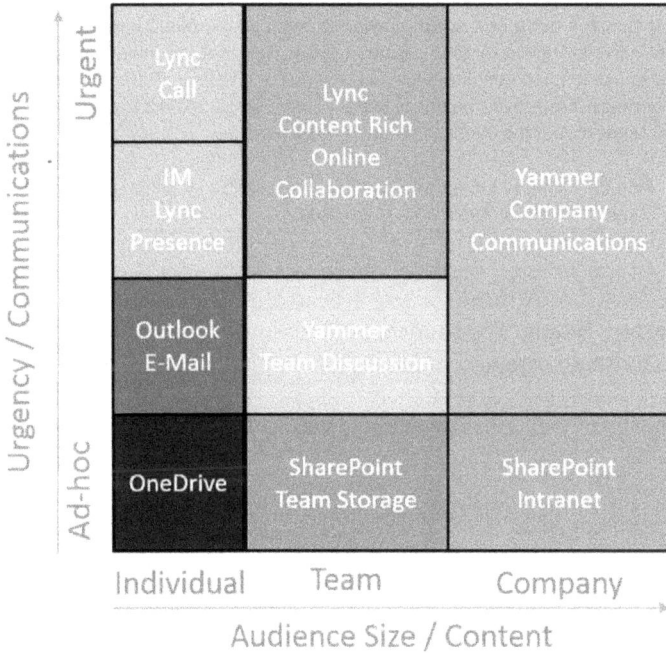

It's important to not only educate users on how to use the technology, but to provide guidance on when and what to use as well.

Authors Note: At Ignite 2015, I will be presenting a session on this topic where I will try to provide more prescriptive guidance for this challenging topic area. Be sure to check http://www.RHarbridge.com and/or Ignite to obtain the latest guidance on when to use what in Office 365.

Seed Content

Suggested Timing: Pre-Launch, Post-Launch, & Ongoing **Level of Effort:** Medium to High

When a user visits a new system, it is important that the system not only contains basic content, but also actionable and useful content that matters to the user.

As an example, if you are rolling out a Yammer network within your organization, you may want to seed content to help others better understand how and why they would use Yammer. "Prime the pump" with real questions and answers and create groups that serve a business purpose.

Example: One organization initially deployed a social network tool and, unfortunately, the most popular content was discussions of celebrities and gossip. This is not an effective or useful network for a business, and consequently they failed to get adoption.

When they rolled out a new social network to replace the failed one, they made sure that not only did they have good seeded content, but that the initial pilot groups who wanted to use it could define a business purpose and reason. The resulting initial use cases demonstrated strong business benefit and ensured that not only was the starting content relevant, but the initial community was exemplary.

Later, when they opened the network to all users, people who joined the social network clearly understood how it was meant to be used.

Provide Internal Classifieds/Marketplace

Suggested Timing: Pre-Launch, Post-Launch & Ongoing **Level of Effort:** Low to High

Establish an effective way to redistribute business and personal assets or services throughout your organization. Consider incorporating classified sections and market-places into your intranet.

Within a large enterprise, organizational units—such as divisions in a manufacturing company—can see business value in trading and buying equipment from one another. In one large company, this relatively simple solution on SharePoint provided enormous cost savings as equipment was effectively sold or reused within the company.

Of course, there are many other uses for this application! Since it revolves around a social notice for buying or selling, it often goes beyond the sale of physical assets and can even be used for sharing babysitters, neighborhood restaurants, or any other networking information that could be useful for employees.

This can provide one more reason for people to leverage your portal and become more familiar with the benefits of the Office 365 suite or SharePoint.

Provide Specialized Search

Suggested Timing: Pre-Launch, Launch, Post-Launch & Ongoing | **Level of Effort:** Low to High

Both Office 365 and on-premises SharePoint provide many of the essential elements necessary to enable modern enterprise search within an organization. Unfortunately successful enterprise search requires an organization to invest effort into maximizing the usage and relevancy of these elements. When you look at your enterprise search strategy and the behavior of your users, some of the simplest improvements can have some of the most significant impact.

Beyond the basic example of creating a separate search for documents, here are two other popular search patterns that lead to higher usage and adoption:

- Consider creating a search experience that only searches PowerPoints and presentations in your organization. Not only can it help individuals find what they are looking for quickly, but it can act as a great way of showcasing popular or best-bet presentations or templates that should be leveraged. If you enable users to submit their presentations to be featured, you can foster the sharing of work that your users are proud of while facilitating discovery of great content.
- Every organization has acronyms that not every employee is familiar with. Create an Acronym Wiki or Acronym Finder that provides users with the acronym explanation as well as relevant documents, teams, contacts and more within the company.

Provide Usability Testing, Diary Studies, Card Sorting Exercises & More

Suggested Timing: Pre-Launch, Launch, Post-Launch & Ongoing | **Level of Effort:** Low to High

Providing lots of design, information architecture and feedback activities can greatly improve adoption by enabling you to better understanding your users, their behaviors, and their challenges with your design, and can furnish opportunities to improve the system.

Develop Communities & User Groups

Suggested Timing: Pre-Launch, Launch, Post-Launch & Ongoing | **Level of Effort:** Low to Medium

Help people connect and sponsor or support the sharing of business practices, information and ideas in your organization by creating and engaging with internal communities and user groups.

At a minimum, fostering communities and user groups is a great way to help people connect and better understand the technology, while providing a channel for feedback and enterprise engagement. In many cases, it is better for the organization to invest in and develop these communities and user groups (champions are your strongest adoption asset).

Provide Internal Content Services Support

Suggested Timing: Pre-Launch, Launch, Post-Launch & Ongoing | **Level of Effort:** Medium to High

One of the most requested areas of support is help with planning and providing effective content. This can include providing support for migrating content, permissions planning, or information architecture planning. Consider developing engagement models internally where you can go beyond basic education services and provide support in getting a stakeholder's content into their site or group, structuring appropriate permissions and rights protection, and determining the best way to ensure search, sharing and management of content is as effective as possible.

Provide Business Intelligence (BI) Services and a BI Community

Suggested Timing: Pre-Launch, Launch, Post-Launch & Ongoing | **Level of Effort:** Low to High

Many organizations are undergoing a digital transformation and becoming a data-driven culture. To support this, consider the creation of a BI @ Company site that provides a single place to explain:

- The value of Power BI
- Benefits of the data catalog and what data sources are available
- The importance of BI to the company
- How people can obtain mentoring or BI support services
- How people can request access to more data or data sources

Recommended Next Steps

The next steps that are most applicable to you might depend on the status of your Office 365 rollout.

Have You Already Launched?

Don't worry, there is still plenty you can do to increase and improve adoption:

- Take a look at the Pre-Launch and Launch activities.
  - Have you done all of them?
  - If there are some you haven't done, consider implementing them and proactively announcing and spreading awareness of the new resources to your existing Office 365 users.
- Plan and begin executing many of the Post-Launch and Ongoing activities.

It's never too late to get a baseline understanding of where your adoption is today and where you want it to be.

- Struggling with prioritizing or determining the best way to do some of the activities described in this chapter?
  - Reach out to a trusted adoption expert who can help you, and connect with other customers to learn what they have done and how to avoid some of their mistakes.
  - Be sure to look into existing products that may already provide the additional functionality or reporting that you require to be successful.

Are You Planning Your Launch?
It's critical that you create an adoption roadmap that outlines which activities you plan on accomplishing and how you will measure milestones and success for adoption of Office 365.

- Develop an adoption roadmap. The roadmap contains activities described in this chapter, milestones and critical dependencies.
- Plan and begin executing many of the Pre-Launch activities.
- Be sure to establish an effective baseline of adoption and track progress on an ongoing basis as the organization and the related initiatives change.
- Research and connect with other customers to see how they have driven and guided adoption, so that you can learn from their successes.
- Consider getting an expert's help.

As with any good Office 365 deployment, feedback is critical to our continued success in providing the guidance you need. So please don't hesitate to provide feedback on this book to the Microsoft adoption and success teams, or to the author of this chapter (who adores your brilliant and candid feedback), Richard Harbridge (Richard@RHarbridge.com).

Appendix Content

Adoption Activity Reading Checklist

What follows is an adoption activity checklist you can use to track which activities you have completed or which ones you plan on completing as you improve adoption within your enterprise. The activities are listed according to category, along with their level of effort (LOE). Once you have identified which activities are relevant to you, ensure they are outlined effectively in an adoption roadmap so that you can coordinate across your team and organization.

Pre-Launch

Before you launch Office 365 or the latest on-premises server technology, you should consider each of these activities.

Leadership	LOE
Identify Key Stakeholders	L
Engage Executive Sponsors	L
Create a Well-Defined Vision for Office 365	L
Determine Actionable Goals & Map to Solutions In Office 365	M-H
Define Your Metrics for Success	M-H
Create Your Communication Plan	L
Train Your Helpdesk	L-M
Awareness	
Leverage Champions	L
Create Countdown Email Campaigns	L
Broadcast Portal Announcements	L
Post Internal Advertisements	L
Show Teaser Videos	L-M
Stage an Awareness Event	L
Create Real Organizational Profiles	L
Run "For Charity" Reward Campaigns	L
Incorporate Feedback in Portal	L
Include an FAQ Listing & FAQ Finder	M

Education	
Create Pilot & Early Adopter Yammer Groups	L
Create Training Site	L
Hold In-Person Training Events	L-H
Conduct Baseline Survey on Office 365 Understanding (Champions)	L
Conduct Champion Experiences Survey	L
Communicate Policies & Guidelines	L-H
Provide Business Intelligence (BI) Services and a BI Community	L-H
Seed Content	M-H
Provide Usability Testing, Diary Studies, Card Sorting Exercises & More	L-H
Provide Internal Content Services Support	M-H
Develop Communities & User Groups	L-M
Hold YamJams, Use #YamWin & Create Starter Yammer Groups	L-M
Provide Internal Classifieds/Marketplace	L-H
Provide Specialized Search	L-H

Launch

As you launch Office 365 or the latest on-premises server technologies, you should consider these activities.

Awareness	LOE
Send Announcement Emails	L
Host a Launch Event	L-H
Conduct a Baseline Survey & Follow-Up	M
Create a Feedback & Improvement Yammer Group	L
Hold Contests & Competitions	L
Run "For Charity" Reward Campaigns	L
Incorporate Feedback in Portal	L
Remove Alternatives & Make It Mandatory	L
Education	
Hold YamJams, Use #YamWin & Create Starter Yammer Groups	L-M
Provide Best-in-Class & Solution Showcases	M-H
Provide Guidance on What Tools to Use for What Purpose	M-H
Provide Specialized Search	L-H
Provide Usability Testing, Diary Studies, Card Sorting Exercises & More	L-H
Develop Communities & User Groups	L-M
Provide Internal Content Services Support	M-H
Provide Business Intelligence (BI) Services and a BI Community	L-H

IMPROVE IT!

Post-Launch & Ongoing

After Office 365, an Office 365 workload, or the latest on-premises server technology is deployed and available, you should consider ongoing activities like these to further bolster adoption.

Awareness	LOE
Send Weekly or Biweekly Tips & Tricks Emails	L
Schedule Buzz Days (Continual Awareness Events)	L-M
Conduct User Experience Surveys	L
Showcase Spotlight Days	L-M
Report on Success Metrics	L-M
Incorporate Engaging Content into Your Portal	L
Incorporate Daily Performance Metrics into Your Portal	L
Create Mock Profiles	L
Create Real Organizational Profiles	L
Incorporate Profile Links in Employee Signatures	L
Hold Contests & Competitions	L
Run "For Charity" Reward Campaigns	L
Define Portal & Site Identities	L
Create Baited Email Hooks	L
Incorporate Feedback in Portal	L
Remove Alternatives & Make It Mandatory	L
Education	
Share Summaries of Top Searches & Corrected Failed/Abandoned Searches	L-M
Review Usage Reports & Relevant Audited Activity	L-M
Connect, Teach & Empower Champions	L-H
Hold In-Person Training Events	L-H
Incorporate Portal into New-Hire Orientation	L-M
Incorporate Portal into Annual Training or Skills Validation	L-M
Announce & Explain New Features & Capabilities (Office 365 Update Announcements)	L-M
Communicate Policies & Guidelines	L-H
Collect, Share & Support Success Stories	L-H
Hold YamJams, Use #YamWin & Create Starter Yammer Groups	L-M
Provide Best-in-Class & Solution Showcases	M-H
Provide Community Showcases	L

	Provide Guidance on What Tools to Use for What Purpose	M-H
	Seed Content	M-H
	Provide Internal Classifieds/Marketplace	L-H
	Provide Specialized Search	L-H
	Provide Usability Testing, Diary Studies, Card Sorting Exercises & More	L-H
	Develop Communities & User Groups	L-M
	Provide Internal Content Services Support	M-H
	Provide Business Intelligence (BI) Services & a BI Community	L-H

NOTE

[1] *Kegan, R., and Laskow Lahey, L. (2009). Immunity to Change. How to Overcome It and Unlock the Potential in Yourself and Your Organization. Boston: Harvard Business Press. ISBN 978-1-4221-1736-1. http://www.amazon.com/Immunity-Change-Potential-Organization-Leadership/dp/1422117367*

ABOUT THE AUTHOR

Richard Harbridge is an author and internationally recognized expert in Office 365, SharePoint and Collaboration. He has defined, architected, developed and implemented hundreds of Office 365, SharePoint, and Azure solutions for customers around the world.

As a sought-after speaker he often shares his insights, experiences, and advice around collaboration, knowledge management, social computing, ROI, governance, user adoption, and training at many industry events around the world.

When not speaking at industry events Richard works as the CTO of 2toLead helping Microsoft, Microsoft partners, and Microsoft customers as an advisor around business and technology.

He has been involved in the industry since the beginning of his career and is an excellent (and friendly) contact on anything relating to Microsoft, Office 365, Enterprise Content Management, and Social Networking.

Chapter 5

The Road to Awesome Adoption

A Successful SharePoint Rollout
Starts with Understanding and Ends with Action

by Heather Newman and Simeon Cathey

My business partners—Simeon Cathey and Mike Thompson—and I have been involved with SharePoint since its birth, 17 years ago. We started our business, Content Panda, four years ago to provide an in-context help content solution that drives adoption of Office 365 and SharePoint. We have worked with many organizations to roll out new deployments of Office 365 and SharePoint, and have also helped enterprises correct unsuccessful deployments. We began to see patterns emerge as customers and colleagues struggled with deployment, training and adoption, and we developed strategies and recommendations based on this first-hand experience.

The best rollouts start with understanding a few key elements and then developing clear action plans. Implementations can be technology-driven, but must be led by executives with buy-in and vision. In order to carry out a successful deployment that delivers the business value an enterprise seeks from its investment, you need access to expert help and guidance.

I'd like to share with you the following resources and guidelines that can help your organization achieve a successful rollout of Office 365 or Share-Point and lead to maximum adoption and usage across your business.

The Journey Starts Here

To guide your journey on the road to awesome adoption, I have organized these recommendations based on a few guiding principles. Follow these principles as you plan your trip:

- Understand the "human factor."
- Understand your why.
- Use an adoption checklist to take action.
- Develop your adoption toolbox.
- Seek out new methodologies to drive faster adoption.

Understand the Human Factor: Driving Adoption Is About Human Beings, Not Technology

A great colleague of ours, Bryan Kramer of Pure Matter, wrote a terrific book called *There Is No B2B or B2C. Human to Human: #H2H*. I love what his approach represents for marketing, but I also think it applies to software and business. We talk about this concept a great deal at Content Panda, and this is the core of his message:

> Businesses do not have emotion. People do.
> People want to be a part of something bigger than themselves.
> People want to feel something.
> People want to be included.
> People want to understand.
>
> Communication shouldn't be complicated. It should just be genuine and simple, with the humility and understanding that we're all multi-dimensional humans, every one of which has spent time in both the dark and delightful parts of life.
>
> That's human to human. That is #H2H

Human beings are innately complex yet strive for simplicity. Our challenge as humans is to find, understand and explain the complex in its most simplistic form.[1]

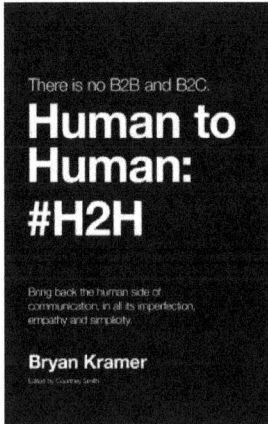

There is no B2B and B2C.

Human to Human: #H2H

Bring back the human side of communication, in all its imperfection, empathy and simplicity.

Bryan Kramer

Edited by Courtney Smith

Technology should be simple, easy to use, and make your life less complicated, not more. Remember that people are the "social" in social technology. Make it easy, fun and rewarding for them to use.

The human factor is the essential element throughout your deployment plans. So as I share the other recommendations to drive adoption and a successful implementation, I will point out the human factor in each one as guideposts to aid you in your journey.

Understand Your Why: Define Your Vision First

The place to start in any implementation of technology is with your "why." Determine your objectives before you begin, to make sure the project is one that everyone is excited about and ensure success.

Sometimes technology is brought in and imposed on an organization by someone without investigating why or if it's the right solution. Sometimes it may even be implemented for the wrong reasons. Occasionally the technology has been sitting there unused and IT decides to turn it on simply because it's already there.

If you are implementing a technology solution, stop and ask the questions, do the research, investigate. What are our business priorities? Who made the decision to bring in this technology? What are we going to use it for? What is our implementation time frame? How do we measure success? Determine the why. Then you can develop the vision of what a successful implementation looks like and the business value you hope to achieve.

Simon Sinek, in his book *Start with Why: How Great Leaders Inspire Everyone to Take Action*, has some very insightful things to say on the subject:

> Very few people or companies can clearly articulate WHY they do WHAT they do. By WHY I mean your purpose, cause or belief — WHY does your company exist? WHY do you get out of bed every morning? And WHY should anyone care?
>
> People don't buy WHAT you do, they buy WHY you do it.
>
> We are drawn to leaders and organizations that are good at communicating what they believe. Their ability to make us feel like we belong, to make us feel special, safe and not alone is part of what gives them the ability to inspire us.
>
> Studies show that over 80 percent of Americans do not have their dream job. If more knew how to build organizations that inspire, we could live in a world in which that statistic was the reverse - a world in which over 80 percent of people loved their jobs. People who love going to work are more productive and more creative. They go home happier and have happier families. They treat their colleagues and clients and customers better. Inspired employees make for stronger companies and stronger economies.[2]

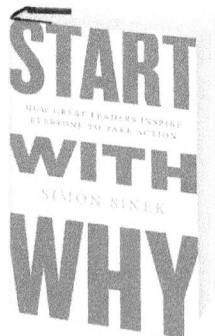

Human Factor: Visioning is often done as an afterthought. What if we lead with it? Defining our why, our vision, gets people on the same page. There is nothing worse than going down a path only to find misconceptions in play or expectations that were never defined. Start with vision. With this vision you can chart a course that people will follow.

Tick the Boxes of Your Adoption Checklist

According to the 2015 AIIM Industry Watch, a failure of senior management to endorse and enforce SharePoint was the biggest reason for lack of success, followed by inadequate user training and a general lack of planning. In their survey, 26 percent of respondents reported that their SharePoint project had stalled and 37 percent have struggled to meet their original expectations, meaning a total of 63 percent had suboptimum installations.[3]

Adoption checklists have been around for long time and yet most organizations do not use them. Driving adoption takes time, energy, and appropriate methodologies and strategy choices. Here is a basic list of essential to-dos that we have used and seen in hundreds of variations in the marketplace. Use them to help draft your adoption checklist:

Secure Executive Buy-In

Your adoption plan should include an executive sponsor with an internal communications plan. All successful deployments have endorsement by the C- suite. This means that the leaders have bought off on the spend, understand the ROI, have agreed on metrics and will be spokespeople and power users of the technology. Then loop in your corporate communications person or team to create persuasive launch emails, develop blogs explaining the "why" and maintain a continued presence with check-ins, updates and tips that extend visibility and support way beyond the kickoff. A best practice is to set up an internal blog for your CEO or executive sponsor on SharePoint. Send the lead paragraph in email to the entire company, driving people to your corporate portal; then require them to go to the site for pertinent and critical information.

Human Factor: People follow their leaders. When endorsement comes from the top, it means the business has taken the time to listen and take part. Then everyone else can feel safe—and will be inspired—to join them.

Create a Pilot and Feedback Team

Start small with a group that will become the guinea pigs: marketing, human resources or the product documentation team. Roll out a divisional site, and provide an in-context training app that will bring training and help content directly into the software. There are many for you to choose from—and quite a few are free (enter shameless plug for Content Panda). Gather initial feedback; see what is working and what is not.

Human Factor: By gathering feedback you are listening to the business and starting to build supporters of the initiative.

Use the Pilot and Feedback to Identify Your Use Cases

What are the most important business scenarios for your organization's success that you would like to see the new technology positively impact? Look to the experts for guidance and examples. Microsoft has put together a booklet and website identifying the top use cases and scenarios demonstrating the many ways SharePoint can help you work better together—on both the enterprise and divisional level.[4] The key here is to roll out slowly while capturing feedback and homing in on what will ultimately drive productivity and revenue for your business.

Human Factor: Lean on best practices of those who have done this before and choose your use cases wisely.

Identify Your Champions

You have executive support and a small trusted group who helped you roll out the pilot and garner feedback. Now it's time to identify the early adopters who can play key roles in the overall adoption of the software. You can institute a train-the-trainer or lunchtime learning program that becomes a part of your communication rollout and training and support plans. Have these folks own an internal knowledge base and reward them for their efforts. Give them a name, a logo—make it fun for them to be the leaders of your efforts.

Human Factor: Some people love to lead and be recognized for it. Empower these folks to empower the rest of the organization.

Create Your Pre-Launch Plans

Once you've initiated your pilot and identified your champions, now the fun begins: the creation of your launch plans. The following three types of plans are essential elements in this step:

- Governance plan
- Communication rollout
- Training/support plan

In the Office 365 and SharePoint community there are many great resources, and knowledgeable people who have years and years of experience creating effective launch plans. Many of these experts are Microsoft Most Valued Professionals (MVPs), a group of non-Microsoft employees that are awarded the title of MVP and are highly sought-after advisers on various Microsoft products, technologies and services.

Don't feel like you have to start from scratch, because you don't. As you do your research, look to the blogs of these experts; read through their advice and see which methodology feels right to you. Do not skip this step. Here is a short list of MVPs and their specialties:

- **Governance:** Christian Buckley, Sue Hanley, Dan Holme, Eric Riz
- **User adoption:** Robert Bogue, Matthew McDermott, Jennifer Ann Mason, Asif Rehmani, Christina Wheeler, Richard Harbridge

What is a governance plan? Governance is the set of policies, roles, responsibilities and processes that control how an organization's business divisions and IT teams work together to meet organizational goals.

What is a communications rollout? This is the plan that informs your business that you are implementing a new technology, and explains how it works, how it will roll out, what the timeframe is, who is on the team and how everyone is expected to participate.

What is a training/support plan? This is the plan of how you will roll out training and education of your teams. Identify who will own this effort. This plan should also include plans for long-term support. Refresh materials as new information comes out. Most people will not remember everything they learn from an initial look at the technology. Make available FAQs, wikis, and additional guides and resources to reinforce what has been learned and help with overall retention and effectiveness of training materials.

Human Factor: "If you fail to plan, you are planning to fail!" Our founding-father of business planning advice, Benjamin Franklin, was once again on the money (which is probably why he *is* on the money) with this quote. The more you do up front, the more successful you will be. This is hard work and takes time and dedication. Write those plans.

What's in Your Adoption Toolbox?

Training is usually the basic foundation of an adoption toolbox. There are many types of training available today. They include live instructor remote training, on-demand instructor recorded training, self-guided training, peer-to-peer sharing, and technical support. What you choose will depend on what kind of budget you have and how you have created the training/support plans described above.

Encouragement and gamification are other great ways to continue expanding the success of your deployment. There should be excitement and buzz now around the new system, and to keep the momentum going you need to reinforce it. Try assigning teams certain tasks or giving rewards to first timers. These tactics can help you gain insights into how people use the software and capture best practices for daily activities, and can build productivity and positivity in the business.

Human Factor: Initial training should be rolled out in small dose and then followed up by resources, encouragement and rewards. Make the adoption of Office 365/SharePoint something that the business is proud of and can brag about. Remember that people tend to forget how to do things and they often will struggle instead of asking for help. Provide ongoing support and guidance.

Integrate New Methodologies & Apps

In-context and just-in-time training app options, which bring help content inside the software, are valuable aids that are becoming more and more popular today. Context-sensitive, or just-in-time, help is a kind of online help that is obtained for a specific point or part of the software, providing help for situations associated with that feature. This context-sensitive help, as opposed to general online help or online manuals, doesn't need to be accessible for reading as a whole. Each topic provides extensive descriptions and help for one state, situation or feature of the software.

Context-sensitive help can be implemented using tooltips, which provide a concise description of a GUI widget or display a complete topic from the help file. Other commonly used ways to access context-sensitive help are by clicking a button or a widget.

Content Panda, for example, provides a new and easy way for end users to access help content. Instead of users going outside the software to hunt and peck for "how do I...?" in a search engine, we bring curated content directly into Office 365. Targets light up giving end users the information they need at their fingertips. The help content that come out of the box in Office 365/SharePoint is good. However, mapping content directly onto the GUI elements elevates the experience so that end users feel empowered and can self-serve most basic support questions and issues. Content Panda is a free app available to all for Office 365 and SharePoint 2013.

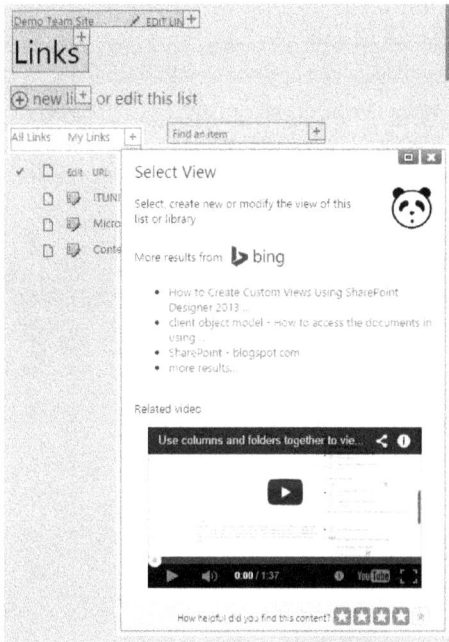

Human Factor: In-context apps give users an easy way to get the help content, or just that simple reminder they need to complete a basic task. Empower people to help themselves and they are instantly more productive.

Conclusion: The Road to Awesome Adoption

Driving adoption takes time, planning and buy-in. Whenever new technology is brought into an organization, remember it's the people that make or break it, adopt it or ignore it. People are resistant to change. So pay attention to the human factors in social technology. Start with your why and formulate your vision. Develop and use an adoption checklist. Build your plan leveraging guidance from the abundance of highly successful Office 365/SharePoint MVP advisers available. Research what training is available within your budget. And implement one of the free productivity apps to empower your organization and help you get the most out of your software investment. Follow this roadmap and help steer your company to successful adoption. Happy travelling!

NOTES

[1] *Bryan Kramer. There Is No B2B or B2C. Human to Human: #H2H. http://www.bryankramer. com/there-is-no-more-b2b-or-b2c-its-human-to-human-h2h/*

[2] *Simon Sinek. Start with Why: How Great Leaders Inspire Everyone to Take Action. http://www. amazon.com/Start-Why-Leaders-Inspire-Everyone/dp/1591846447.*

[3] *2015 AIIM Industry Watch Report. Connecting & Optimizing SharePoint - Important Strategy Choices. http://info.aiim.org/2015-sharepoint-report*

[4] *Discover SharePoint. http://aka.ms/spuc. http://success.office.com/.*

ABOUT THE AUTHORS

Heather Newman is a technology entrepreneur and award-winning senior marketing professional that thrives in both startups and established corporations. Her experience in building global high-tech marketing strategy (B2B space) has helped drive revenue for companies through alignment with sales, partner channel, leadership teams and clear execution of initiatives. She has led global marketing organizations for two well-respected, award winning and best in class Microsoft partners, AvePoint & KnowledgeLake. Heather was a part of the original SharePoint Marketing team where she produced thousands of events for Microsoft including the first three SharePoint Conferences. Heather serves on the board of AIIM International, is an Ambassador for The CMO Club and is Co-Founder/CMO of Content Panda and CEO/Head Maven of Creative Maven. She can be found online at www.creativemaven.com and @heddanewman.

Simeon Cathey is a technology visionary, entrepreneur and Microsoft expert with more than 15 years of experience building information and document management applications for businesses. Simeon began at Microsoft as part of the original SharePoint team in 1998, and served as the SharePoint Product Release Manager for Microsoft SharePoint 2003 where he was at the forefront of the enterprise knowledge management and collaboration movement as it began to take shape. Simeon is Co-Founder and CEO of Content Panda. He is also a published author and thought leader, having written Wiley Publishing's Microsoft SharePoint Server 2007 Bible. Simeon also serves on the board of Tis Best Philanthropy. He can be found online at www.simeoncathey.com and @simeoncathey.

Chapter 6

Enterprise Search

Planting and Cultivating to Reap Long-Term Benefits

by Agnes Molnar

Deploying *enterprise search* is not a single IT process; it's much more complex than that. Actually, enterprise search should be considered an enterprise business process. The stakeholders have to understand how it works, but at the same time they must identify the expectations and requirements, too. This includes addressing where the results come from, what kind of results can be expected, and what the user experience should be.

Like a garden, SharePoint search is an intricate system composed of many functional components that interact organically. After planting the seed of implementation, a company grows, user's needs change, and the search system requires ongoing care and maintenance.

Even the simplest search solution is a complex element of the company's information architecture.

Common Challenges

The first and perhaps the biggest challenge is a misunderstanding. People often consider search to be easy. Their impression is that search just has to be "turned on" and it works. Technically speaking, this might be true; but the problem is always with the results it provides: Are these results relevant to the user? In today's insane information overload, the biggest challenge is with *findability* and *discoverability* of the information. Getting millions of results doesn't make sense in the enterprise, but we do want to get *all* the relevant documents.

The next challenge is how to define search requirements. "I just want it to work" is not a requirement. Neither is "I want a nice, fancy search" nor "We want to be able to find everything."

When defining the *formal* requirements, we have to focus on three aspects of search:

> *Content:* It's obvious that we need content to be included in search. This includes types of documents and objects, and their structure, attributes, and meta-information.

> *Users:* Every user has different information needs and a unique personality. We have to identify the audience types, users' expertise, and typical tasks.

> *Context:* Last but not least, in order to define the global requirements of search, we must understand where and how the information is used.

When specifying these requirements, we have to plan not only for the implementation of search, but also for its maintenance and metrics. To be able to make and keep search successful, we have to identify our success criteria as well.

What Makes a Good Enterprise Search?

This is an important question.

Because every user has different information needs, and our needs might be different from query to query, defining what a *good* search means might also vary from person to person. The key point is to find the common needs.

One of the most popular metrics is *search usage*. It rests on a very simple observation: even if people start using search, they stop using it if they don't get the desired results. Checking the usage analytics right after the release is a necessary step, but is definitely not enough. We need to keep checking these reports, analyze them, and take actions when and where needed.

Besides usability, security is critical, too. Although SharePoint search always provides the results security trimmed, it's important to understand what this actually means:

- Security trimming means users cannot see objects as search results that they don't have authorized access to.
- With systems than can be connected by out-of-the-box connectors (like SharePoint, file shares, websites, Exchange public folders, etc.), security trimming is given; we cannot override or "hack" it. With custom connectors though, we have to take care of the proper mapping of the source system's permissions into SharePoint access control lists (ACLs).
- Security trimming is a strong and useful feature. Users can find the content that is indexed *and* that they have access to. However, problems can result if the content source's permission settings are not maintained properly.

Here is an example of problems caused by incorrect permission settings: A financial company had millions of documents in a huge file share with a very deep and wide folder structure. In one of the sub-sub-subfolders, they had a file with the name "ManagementSalaries.xls". As you might think, this file contained all the managers and C-level executives, with their salaries, cafeterias and other company benefits. Since it was very deep in the folder structure, almost nobody knew that it existed.

But as soon as we added this file share to search and indexed its content, this Excel file started to appear in the results, actually in a very high position when someone was searching for his or her manager's name.

Of course, this was a huge security problem. The company didn't want to disclosure the content of this document. But it's important to clarify that even though it happened in SharePoint search, the problem was on the content side. If the file share's permissions had been set correctly, employees could not have found this document.

In some cases, there is a need to be aware of every piece of content that does exist, regardless of its access rights. In this circumstance, users can apply for read permissions instantly. Due to the way the SharePoint search engine works, it's impossible to "hack" the search engine itself to support this kind of behavior. Instead, we always have to provide a custom solution that is either based on a custom crawler or a general content "inventory."

Due to the complexity of content sources and search itself, it's always necessary to run thorough and deep security tests before making search generally available to everyone. These tests must be planned and executed in an accurate way.

Maintaining and Preserving Your Search Garden

While planning and implementing search, keep in mind that it should be governed and maintained, even long after implementation is complete. We usually say that search itself never can be "done." Consider it like gardening. You set up your garden, plant the trees and flowers, water them and enjoy a beautiful first blossoming. But the work hasn't ended at all. If you don't water it regularly, if you don't prune the trees, fertilize and weed the garden, or mow the lawn, your lovely garden can rapidly turn into a barren field or a chaotic jungle.

Enterprise search is very similar to this. You don't allocate the proper resources, don't take care of it, neglect maintenance and updates...and it gets messy very soon. The results lose their relevancy, users don't get the experience they want; and then they stop using it. Don't forget: as the environment changes around us, so does your business. User needs evolve, and search has to follow them in order to be and stay successful.

Of course, to be able to keep this up, we need a team. Depending on the size of your company and your needs, the size of this team can vary. The first important point is to realize the need: you have to allocate resources for these tasks. Then you can determine if it's going to be one person or ten.

Table 1 contains some of the typical internal and external search roles — not only during the implementation phase, but also later, during maintenance and support.

Table 1

Search Roles	
Internal	**External**
• Business Sponsors	• Consultants
• Stakeholders	• Service Providers
• Project Management	• Product Vendors
• IT / Support	• External Developers
• Search Admins	
• Internal Developers	

Sometimes it's also necessary to have someone in a liaison role. This person acts internally to represent the interests of the customer, and must understand the needs and business motivations. At the same time, he or she has to have a deep understanding of enterprise search and be able to serve as liaison between the customer and external team members (in most cases, consultants and external developers).

Now that we've covered the basics, let's see how to measure the success of any enterprise search implementation. The appropriate metrics have to be identified and defined in advance. Designate the team members who have responsibility to analyze each of these metrics regularly, create reports and decide on follow-up action plans.

Figure 1 shows some of the most commonly used metrics. But of course, this list is not (and cannot be) complete. Every business has to define its own success factors and metrics during the planning phase.

Figure 1. Common Enterprise Search Metrics

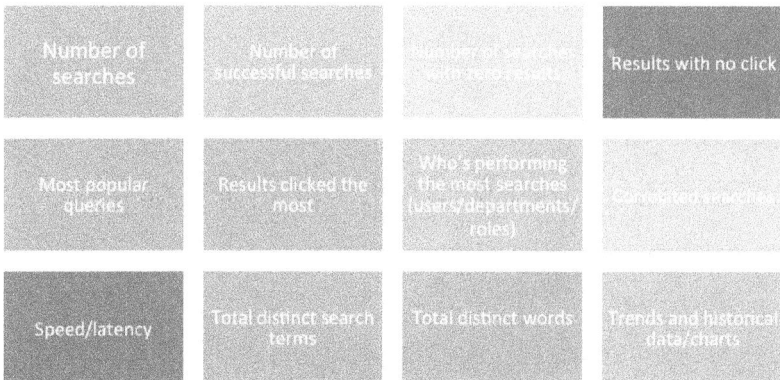

Figure 1. Common Enterprise Search Metrics

Number of searches	Number of successful searches	Number of searches with zero results	Results with no click
Most popular queries	Results clicked the most	Who's performing the most searches (users/departments/ roles)	Convoluted searches
Speed/latency	Total distinct search terms	Total distinct words	Trends and historical data/charts

When we have the plan ready, the team set up and all the relevant metrics defined, it's time to take action. As I mentioned at the beginning of this chapter, search is a continuous business process, like gardening, rather than a one-time project. Therefore the management and action plans must reflect this long-range approach, too.

If there's an existing search solution in your business, analyze it, collect feedback and listen to "user voice." Crawl reports can provide valuable information about users' behavior, too. Also, check the logs for any existing and permanent errors and fix them. There are many tools to extend Share-Point's out-of-the-box reporting and analytics capabilities — it's time to use them!

User experience and metadata definitions must be reviewed and improved on an ongoing basis, too.

As Figure 2 illustrates, this "search gardening" process is a continuous cycle of planning, implementation and analysis.

Figure 2. Search Maintenance Cycle

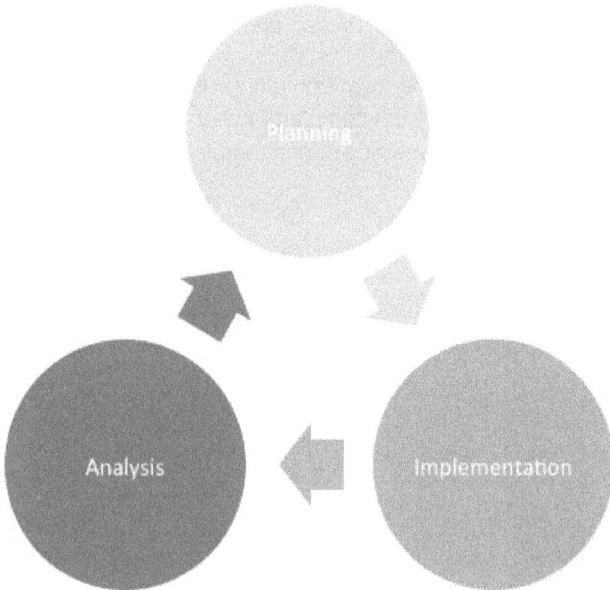

As we've seen in this chapter, *proving* the value of search is not enough — we always have to work on *improving* it. There are many tools and techniques to do so. But first we have to recognize and remember: search is a business process, not a one-time implementation project.

To keep our enterprise search fruitful and productive, let's get gardening!

ABOUT THE AUTHOR

Agnes Molnar is CEO and Managing Consultant of Search Explained, specializing in Information Architecture and Enterprise Search. She started her career as a developer while she was studying Information Technology at the Budapest University of Technology and Economics. By the time she acquired her Master's Degree, she has already established a solid profession in Computer Science. She has worked for various companies in Europe and the US, developing dozens of SharePoint and FAST implementations for both commercial and government organizations throughout the world. Her familiarity and extensive knowledge in SharePoint and other Content Management Systems enabled her to start her own consultancy business, and to eventually become an Independent Consultant. Since 2008, Agnes has been a consistent recipient of the Microsoft Most Valuable Professional (MVP) Award, for actively sharing her technical expertise. She is a regular speaker at technical conferences and symposiums around the globe. She has also co-authored several books and white papers. She also maintains her passion and dedication on the subject through her blog, www.aghy.hu, where she shares troubleshooting tips, best practices, and other useful resources in Content Management with a light and wholesome approach.

Chapter 7

He Who Controls the Data Controls the Company

*How Politics and Governance Make or Break
Analytics Projects in Hybrid Environments*

by Sadie Van Buren

Why Should You Read This Chapter?

A majority of today's IT portfolios include one or more active business intelligence (BI) or analytics projects. If you're an IT manager, director, VP, or CIO, a technical project manager or consultant, or the leader of a functional area in your business, you are responsible for at least one strategic initiative involving analytics or business intelligence. Most of us know by now that technical projects don't fail because of the technology. However, the buzz around cloud/hybrid environments has put a spotlight on the selection of the underlying platform and services for these projects. While these are critical components, it's important not to forget rule #1 in BI/analytics projects: Most anything is technically possible. It's the internal politics and governance that get in the way.

This has been the case since reporting systems and data warehousing first emerged over 30 years ago. BI solutions often require data from multiple sources that cross organizational boundaries. How can you ensure that you have buy-in from the source organization? Have they bought into your accessing the data? Have they provided the correct subject matter experts?

In addition, once data sources are initially integrated, loading from these sources is an ongoing process for the life of a BI solution (often five to ten years). Source data is never perfect, there are always exceptions (e.g. data may not be consistently formatted, null values may be provided for required fields). Do you have correct data quality processes in place? More importantly, is data governance, or data stewardship, in place to ensure that correct data gets reviewed as well as loaded, and that adjustments are being made for exceptions?

With these issues in mind, I interviewed some of the most senior data architects I know, and asked them to share their experiences about factors that could make or break an analytics project.

"But," you say, "There are tons of good articles already out there about governance of data projects—do I really need another one?"

Agreed, there is a lot of existing guidance about success factors (Align your use cases to the business! Have an executive sponsor! Get buy-in from all stakeholders! Define and document your processes! Ensure data integrity!). And much advice exists on worst practices (Don't get bogged down with minutiae, but don't stay too high-level either! Don't assemble a steering committee without a clear purpose and framework! Don't assume that data governance is a project that will ever end!). You have access to a digital sea of carefully-worded, optimistic information about data governance.

Let the following nuggets of truth be your islands of reality in this sea, built on the experience of senior architects who've spent years in the data trenches. Ask yourself these six questions about your analytics projects, whether they are still in the planning stages or well underway. I've provided red flags to alert you to issues and tips to help you mitigate them. I also included some relevant quotes from the experts' own experiences.

1. Who Owns the Source?
(Translation: Is Your Executive Sponsor Executive Enough?)

As I talked with the data and analytics experts in my network, this concept was stressed first and foremost: You will only be able to execute a project within the umbrella of your sponsor. Having the right people at the table accelerates progress (corollary: not having the right people could result in little or no progress at all). Success stories poured out about forward-thinking executive sponsors who were senior enough to make things happen. These were closely followed by memories of failure, where the project teams didn't have access to the data they needed, and as a result, couldn't make the solution shine.

Ownership of the source data is a critical factor. If one group owns all the data sources and they have the incentive to make things happen, the project will progress. If sources are owned by different groups, this exponentially increases the complexity of the project. Some of the departments in your organization may never have shared real-time data before. They might insist on delivering a static text file on a weekly basis, and your Chief Technology Officer (or equivalent) may not have enough authority to force the issue. Some data projects will only succeed if the executive sponsor is the very top executive in the organization.

Red Flag: If the people funding and supporting the project and the people supplying the data are not the same people, you may have trouble getting that data.

> "Anything we needed that got him closer to his goal, he would sign off, and we had access in 24 hours—new tools, calculations, subject matter experts. In this case everything he needed was in his silo."

Mitigate it: Most BI projects don't have the level of scale or scope to engage the CEO. The sponsor of a Sales and Marketing BI project might be the VP of Sales, whereas for an IT-level project it's the CIO or CTO. But if those roles don't have the authority to break through barriers across the organization, you may need to make sure the CEO is engaged. The CFO could also be a powerful ally, because a data project's ultimate goal is to benefit the company fiscally.

2. Are You Driving an Agenda Instead of Running a Project?

Most of the experts I interviewed had at least one story about an agenda disguised as an analytics project. The scenarios ranged from driving a wedge in the enterprise by replacing the incumbent technology with a preferred platform, to the career ambitions of a single executive, to cold wars over funding and resources.

"Their agenda had nothing to do with the tool and nothing to do with the data."

Obviously all projects, IT or otherwise, have some degree of personal agenda behind them. However, your project will fail if it's primarily agenda-driven and not about creating insights that lead to bottom- or top-line benefit to the company.

Red Flag: If you're not allowed to use a commonly occurring word such as "analytics" or "dashboard" when speaking about the project, you're probably furthering an agenda.

Mitigate it: This is another good time to make sure the CEO and CFO are engaged, and that the business value is clear.

3. Could People Lose Their Jobs as a Result of This New Technology?

One of my sources told me he'd come to expect lots of infighting about who controls and accesses data. "They'll ask you for your specific needs, validate those, and give you a snapshot of the data. They realize that once other people get control of the data, their job is called into question. What if they find data that is wrong, what if they find data that's showing they're not doing their job?"

The human impact of analytics projects should not be underestimated. On a basic political level, there's a control aspect. If a group is being asked to share live data for the first time, this means they're being asked to give up control. They'll no longer be able to massage out the knots before delivering the data. And they may need to answer questions and make changes based on recommendations from the project team. In short, they'll be up against the fact that it is the company's data, not their data. That's an uncomfortable shift in thinking.

Beyond the control issue, the exposure of data frightens people because it could cost them their jobs. You're telling them, "We're digging into all of the data that shows your productivity for as long as you've been keeping track. We have tools to show conclusively what the actual productivity was." For example, in a call center, the data might show that idle times for each caller are higher than they should be. Or it could expose managers not being responsive enough, waiting too long to switch workers from a non-busy pool to a busy pool. The company may be going through a round of tightening or integration and looking for ways to introduce automation or to streamline the number of employees manning the phones. The teams generating the data may not even know the extent of what the numbers will show. As a result, there may be a massive internal scramble to improve the operations and procedures that are generating the data, before it starts to be displayed by the new system. If the teams don't know yet what the new metrics will be, they will try to make their existing metrics look better, even if those metrics don't involve the analytic that's eventually considered critical.

"It is terrifying because they are significantly losing control. They don't even know what this information looks like or what it's going to say about them."

Red Flag: If employees start resorting to digital shredding (for example, deleting their log archives), they're probably terrified of losing their jobs. You may also experience passive-aggressive behavior in response to requests you make on behalf of the project.

Mitigate it: Recognize the human factors at work and don't underestimate them. Maybe even—dare I say it—address these concerns openly and communicate empathetically about the strategic change that's happening in the organization.

4. Has a Similar Project Started and Failed Multiple Times?

If your company has engaged in several failed attempts to launch an analytics solution, you'll face skepticism from all sides. This is the polar opposite of the "over-optimism" situation that is sometimes the cause of project failure, and it should be managed just as closely. When a lot of emotional baggage weighs down your project before it even starts, take a lean engineering approach: Build, test and fail fast; iterate rapidly; and show results early and often. Take a product development approach: Rather than waiting until your analytics solution is "perfect," get a minimally viable product in front of your stakeholders as soon as you can. Then release updates continuously, keeping a functioning version live and accessible at all times. This goes for your governance plan as well as your actual solution—put your straw man up and keep refining as you go.

Red Flag: If the project timeline goes over six months, your likelihood of success is drastically reduced.

Mitigate it: Start small and lean with a several-week Proof of Concept that can evolve into your envisioned solution.

5. Can You Adapt to the Analytics?

Metrics will change as a result of your analytics project. As the new analytics come in, be aware that the people who are able to respond to the analytics are those best able to succeed.

In some cases, however, the metrics and key performance indicators (KPIs) may not be right. The data could be bad, the math could be bad, or the metric may not reflect reality. This underscores the importance of having someone who can evaluate the data and conditions, and make the right decisions.

In one example I heard, a company with high turnover in their customer support function did an analysis to see what kind of people they retained the longest. Training was expensive, so retention was the goal. The data showed that a key component in employee longevity was commute distance: The further away someone lived from the support center, the more short-lived their employment was. The company surmised that they needed to hire closer to the support center. However, that center was located in a downtown financial district, some of the most expensive real estate in the city. Workers from more distant zip codes were more likely to be in legally protected socio-economic groups. So not hiring from those zip codes would effectively have been discriminatory. That was reality, but the analytics only saw zip codes.

This is why it's critical to be cautious in the creation of metrics and KPIs. Are they right for the business? Are they achievable? Are your employees equipped and empowered to support them? And are you ready to help your team and the larger organization adapt? Because analytics have the potential to impact areas like professional development, performance, funding, and compensation.

"As more and more metrics surfaced, some managers adapted, grew, and became directors, because they knew how to motivate their staff into the numbers."

Red Flag: If there are people who think they know better than the computer does once the new analytics start coming in, you may not have the right metrics in place. (Then again, you might!)

Mitigate it: Listen to your team leads, remember the support center example above, and make sure you have the big picture in mind as you evaluate and refine your metrics.

6. Are You Playing the Long Game?

Data changes. It requires ongoing care and feeding. Once you finally get everything up and running, people tend to consider the project complete. Companies almost never budget for the following years and the resources necessary to continually ensure data quality and good source data. If you don't have a dedicated data steward over the lifetime of the solution, there's a good chance it will fail. This may not happen right away, but farther down the road when everyone's attention and resources are focused on different initiatives.

Your data steward:

Ensures that your systems have the right checks in place to find bad data coming in—not only bad data types, but broken taxonomies as well.

Keeps your governance plan and data policies up to date.

Stays on top of changing regulatory and compliance requirements.

Works with vendors to define and enforce requirements pertaining to their services.

Represents the needs of all the groups across the organization.

I once said "the data must flow" during an interview. I didn't get the job.

Red Flag: If the project launches and everyone goes away, the solution may fail when no one is expecting it.

Mitigate it: Budget and plan for a dedicated data steward over the lifetime of the solution.

Conclusion

As technology advances, analytics across hybrid environments become increasingly achievable. But don't neglect political and governance factors as you work toward your vision of better business insight through data. Ask yourself the six questions I've outlined, not only at project inception but at key points throughout the process. They can help you concentrate on the aspects of your project that could mean the difference between failure and success.

Acknowledgement: My thanks to Bill Marriott, Ken Seier, Larry Barnes, and the elusive @TripperDay for their time and input into this chapter.

ABOUT THE AUTHOR

Sadie Van Buren spent ten years designing SharePoint solutions and leading deployments, while stealthily managing the human and cultural issues common to technology projects. She is now the Director of Strategic Alliances at BlueMetal, leveraging her years of technology consulting and marketing to build stronger relationships throughout the partner ecosystem. Sadie has a Bachelor's degree from Wesleyan University and a Certification in Project Management from Boston University, and is a Microsoft Certified Information Technology Professional (MCITP). She is the creator of the SharePoint Maturity Model, and blogs at http://amatterof degree.typepad.com/.

Chapter 8

Measuring the Value
of Enterprise Social Technologies

It's All About That Case!

by Susan Hanley

With apologies to Meghan Trainor, it's not *All About That Bass* when it comes to social. It's all about that *case* — the business case, that is.

McKinsey & Company has been studying social technologies for many years, looking at how companies are using and benefiting from "Web 2.0." In July 2012, the McKinsey Global Institute (MGI) published a report called *The Social Economy: Unlocking Value and Productivity Through Social Technologies.*[1] The analysis focused on four major industries: consumer packaged goods, consumer financial services, professional services, and advanced manufacturing. MGI's estimates suggest that by fully implementing social technologies, companies have an opportunity to raise the productivity of high-skill knowledge workers, including managers and professionals, by 20 to 25 percent. Furthermore, they estimate that these industries could potentially contribute $900 billion to $1.3 trillion in annual value by improving productivity across the value chain. That's a pretty decent case!

The vast majority of the potential value estimated by MGI lies in improving communication inside organizations. They estimate that the average knowledge worker spends about 28 percent of the work week managing email and almost 20 percent looking for internal information or tracking down colleagues who can help with specific problems or tasks. When companies use social software, interactions and messages become searchable content. Having this content available can reduce the time spent finding information or people by as much as 35 percent. While it is difficult to directly translate the time saved to specific ROI, the fact that enterprise social makes it easier to find information means that workers can focus on more meaningful actions and other higher value tasks.

All this potential business impact and value does not happen by magic. To get the benefit of social technologies, organizations need to transform their culture and processes to create an environment of openness and trust. More importantly, however, organizations must establish a clear relationship between the use of social technologies and the business challenges of the organization. In other words, no matter how you look at it, it's all about the [business] case.

From Case to Action: Work Backwards

There is a bit of a chicken and an egg dilemma when you are first getting started. The value goes up when everyone participates, but there are often huge hurdles to overcome to get people to participate. While we could identify many possible barriers, I'd like to focus on two: finding the critical moments of engagement and engaging leaders.

- **Finding the critical moments of engagement** is a bit like planning the route to your vacation destination. First, you have to know where you are going. Then you can figure out the best way to get there. When it comes to getting adoption of social technologies, you want to first understand the business scenarios where social technologies can add value. *Then*, you can plan strategies to achieve that value, or "get there."
- **Engaging leaders** is similar. While we don't always need leaders to roll out new technologies, leaders can be very influential when it

comes to changing culture and validating new processes or ways of working. They are not the only key influencers in our organizations, but they sure are important ones!

Find the Critical Moments of Engagement

One of the common traps that organizations fall into when deploying social collaboration tools is to think we "just collaborate." We don't. We collaborate in the context of a business activity, process or task; we engage with other people to get something done. So if we want to be successful deploying social collaboration technologies, we need to find and integrate those critical moments of engagement.

The challenge is that many social collaboration platforms are deployed as stand-alone environments. This forces staff to step outside their usual technology environment—such as their ERP, CRM, or email application—in order to use the new platform. To me, this doesn't make a lot of sense. The best collaboration tools are those that are embedded in the tools we use every day. We're not there yet in all cases. Some social collaboration platforms still need to be integrated, but collaboration vendors are investing here and it is something we need to improve in our own organizations.

When we look at social software, it's very tempting to focus on features. This is another common approach worth re-evaluating. At the end of the day, it's often hard to distinguish among the various tools based on features alone. It's far better to focus on use cases that help us understand how we can *leverage* the tool and its features. Look at the scenarios in your organization where people collaborate today. Where are those interactions or processes especially challenging? The goal is to look for opportunities to make collaboration better, faster and easier. **These are the critical moments of engagement that we are looking for.** The idea is to change the discussion focus from what the technology *is* and what it does, to *how* you are going to solve a specific business problem.

Let's walk through an example. Every sales organization has challenges onboarding new people—helping them get the information and approaches they need to be productive as quickly as possible. Moreover, sales onboarding is an area that typically has an identified key performance indicator (KPI). It's highly likely your organization is already tracking this metric. To introduce your social tools to a sales manager, start with a conversation about how you can help the manager streamline the onboarding process and the sales training process. Show how you can reduce the overall time required to get a new sales person to the *delivering revenue* stage of the process. This conversation is far more likely to get the sales manager's attention than if you approach that person with "Hey, we've got this new tool that I think you should try." Your focus is on introducing how the technology can positively influence challenging business processes and work patterns that are important to the sales manager. It is very likely that you already have a baseline metric for this process. You will now have a much easier time setting a target goal and measuring the impact of your solution, because you have tied it to an existing and meaningful business use case.

By focusing your approach and metrics plan on critical moments of engagement, you can use existing business metrics to clarify what you are trying to do and what success will look like. This will help avoid the potential pitfalls and risks associated with any new technology—especially one that has the potential to change the entire dynamic of the organization.

Engage Leaders

You can't measure the benefits of a technology that isn't being used. This is why identifying the critical business use cases and finding the critical moments of engagement are so important for enterprise social technology. Accomplishing this helps you provide better baselines and measurement of the business value you are providing. These steps also ensure that what you are measuring (the impact) is worth measuring. Beyond the business case owners you engage, there is one group of users who have a particularly important influence on adoption: your business leaders.

For social collaboration to be successful, leaders have to not only explain the "why," but they also have to demonstrate the "how," the behaviors necessary to make use of the tools. This means that leaders must make the connection between the social tools and how they help the business. The leaders have to demonstrate their commitment by participating. When leaders don't participate, they can inadvertently inhibit collaboration and miss out on leveraging capabilities that could create a competitive advantage.

There are many approaches you can use to help engage leaders. In addition, today's leaders themselves must possess several critical social media skills.[2] If you want your initiative to deliver value, all of the leaders in your organization need to be able to:

- **Create interesting and compelling content.** Work with your leadership team to help them understand how to create authentic posts. For example, start with something simple like asking them to post a comment or two about a travel experience to a field office or something they learned recently. Or ask a leader to create a post requesting contributions from employees on a new initiative.

- **Encourage usage by using tools themselves.** Leaders should be encouraged to "follow" the key influencers in the organization and re-post or comment on their posts. It's really easy to indicate affinity with a "like" and interest with a "follow."

- **Learn how to filter out the noise.** Not just the leaders, but everyone using social tools needs to learn to use the settings of the tools to make sure that the important content is separated from the noise.

- **Help raise awareness in others.** It's great if you can encourage leaders to become advisors to their peers. If your leaders are a little apprehensive about new technologies, a great approach is to create opportunities for new employees to "reverse mentor" senior colleagues.

- **Be proactive when it comes to governance.** Social technology moves quickly. It's critical that organizations balance openness and free exchange with guidelines designed to mitigate the risks of irresponsible use. This is a responsibility shared by leaders and employees. The guidelines should address specific issues that might be unique to your organization (such as sharing client or case information or commenting on products that are subject to government regulation). Guidelines should also include advice about when to, and when not to, engage. I have a favorite website that has a comprehensive collection of social media policies from all kinds of industries all over the world. This is a great place to get ideas if you don't yet have a social media policy in your organization, or to find refresher ideas if you do: http://socialmedia governance.com/policies/.

- **Pay attention to the future of technology.** Leaders not already abreast of new technologies should get coaching on emerging trends, and on which emerging trends are likely to provide the biggest benefit in their organization. CIOs have a great opportunity to get a "seat at the table" and drive change when they are able to provide this type of learning for the business leaders in the organization.

Develop a Balanced Approach

Measurement programs for social technologies require a balanced approach. The programs should include system and business process measures—both quantitative and qualitative. Not all of the useful system measures will be available or easily accessible "out of the box" from your tools. To ensure you have many of these useful system measures, you can take advantage of third-party products to facilitate surfacing your key metrics. Measurement provides insights into behaviors and relationships that are necessary for success. But if the time and energy required to collect the metrics are too great, you won't have the data needed to make decisions and course corrections on a timely basis. Be careful about deciding what to count. Balance the ease of collecting a metric with the insight and value you get from collecting it.

Table 1 presents a scorecard for evaluating social technology metrics from three perspectives:

* Health or vitality of the solution
* Features and capabilities of the solution
* Business value the solution can provide

Use this table to help identify some key questions to ask to measure the business value of your social technology investments and for metrics that you can collect to answer the questions.

Table 1

Perspective	Key Questions	Measures
Health	• Are people using the solution? • How many? • Who (which departments or roles)?	• Number of users with complete profiles (overall and by department) • Number of posts • Number of profile searches • Number of blog entries • Number of likes • Number of replies • Number of replies to messages by people who are not specifically mentioned (to demonstrate engagement that wouldn't be possible if the collaboration took place via email)
	Is usage sustained?	Trends over time for each of the key measures above
	What features are used the most?	Tracking and comparison of features such as blog posts, activity posts, likes, and replies

Capabilities	Is usage supporting the identified business use cases?	"Serious Anecdotes"—stories from user surveys where users report specific use cases and value measures based on the moments of engagement identified in the deployment plan
	Do users perceive that they are getting value?	• Survey questions asking users if they feel that they can collaborate more easily and resolve issues more quickly • Survey questions asking whether users can find people with the expertise that they need • Survey questions asking users to rate whether they would like the tool removed (what I like to call the "Don't Take it Away" metric)

Business Value	Is there a clear connection with respect to the overall business strategy?	• The impact on key performance metrics since the social tools have been deployed
		  • Average time for call centers to resolve customer issues
		  • Average time-to-market for new products
		  • Average proposal response time
		  • Average "time to talent" for new employees (cost/time for onboarding)
		  • Annual staff turnover
		  • Customer satisfaction
		  • Ability to handle "exceptions"—situations that don't fit standard processes and require reaching out to experts or multiple departments for resolution
		  • Most popular / most used content. (Be sure to find a way to recognize users who create highly reused or "liked" content and people who benefit or create new content based on content created by others. This will demonstrate that both content creation and content reuse behaviors are important.)

Come Back to the Case

Measurement is critical for enterprise social technology. Adoption metrics alone, such as the number of activities or the number of completed profiles, are not enough to deliver meaningful business results. Meaningful results are demonstrated by business use case measures that identify the critical moments of engagement where social technology is positioned to solve a meaningful business problem.

When social technologies are relevant to the critical moments of engagement in our work, these technologies become tools to get work done better, faster and easier. This provides a critical motivating factor to get users to engage. When you incorporate leadership involvement, the motivating factors are multiplied. Measurement can inform investment decisions, but it also does much more. Measurement can contribute to improved user experiences and allows organizations to be more responsive to change in dynamic environments. If you get this right, you will see that when it comes to social measurement, it's all about the case!

NOTES
[1] *Michael Chui, James Manyika, Jacques Bughin, Richard Dobbs, Charles Roxburgh, Hugo Sarrazin, Geoffrey Sands and Magdalena Westergren. "The Social Economy: Unlocking Value and Productivity Through Social Technologies." July 2012. McKinsey Global Institute. <http://www.mckinsey.com/insights/high_tech_telecoms_internet/the_social_economy>*

[2] *Roland Deiser and Sylvain Newton. "Six Social-Media Skills Every Leader Needs." McKinsey Quarterly. February 2013. <http://www.mckinsey.com/insights/high_tech_telecoms_internet/six_social-media_skills_every_leader_needs>*

ABOUT THE AUTHOR

Sue Hanley is consultant specializing in the "people side" of collaboration and knowledge management solutions. She is the co-author of three popular SharePoint books that focus on business challenges - *Essential SharePoint 2007*, *Essential SharePoint 2010*, and *Essential SharePoint 2013:* *Practical Guidance for Measurable Business Results.* She also writes the Essential SharePoint blog for NetworkWorld at http://www.networkworld.com/blog/essential-sharepoint. Her areas of expertise include strategy, information architecture, user adoption, governance and business value metrics. Immediately prior to establishing her own consulting practice, Sue led the Portals, Collaboration, and Content Management practice for Dell Professional Services. Sue is an Office 365 MVP.

Contact Information For Susan Hanley
Website: www.susanhanley.com
Twitter: @susanhanley
LinkedIn: www.linkedin.com/in/susanhanley
Phone: 301-469-0770
Email: sue@susanhanley.com
Blog: www.networkworld.com/blog/essential-sharepoint

Chapter 9

Expanding Our Collaboration Footprint Through Social

by Christian Buckley

Within the SharePoint community, social tools came to the forefront of the conversation with SharePoint 2010 — primarily citing the lack of social capabilities within the platform. The consumer options available to end users within the enterprise had been multiplying rapidly and the "Facebook Era" was coming up to speed. Organizations were realizing that their employees were not adopting the solutions placed in front of them — or even if they were logging in and uploading documents on occasion, they were not fully engaged. As feedback began to flow into Microsoft, the people had spoken — they wanted stronger social capabilities.

Of course, Microsoft focused a large segment of their strategy on native social capabilities for the next version, SharePoint 2013. But Microsoft found itself in a growing competitive situation with industry-leader Yammer, which led to a $1.2 billion acquisition of Yammer in 2012. There was initially some functional overlap, and some resulting confusion in the marketplace about Microsoft's long-term strategy. However, the company did gain a number of things out of this acquisition: around 4 million registered users, a purely cloud-based platform option, and a strong metric-driven development ethic that employed constant A/B testing.

Out-of-the-box SharePoint provides a very compelling set of features for content management and team collaboration—whether on-premises or in the cloud. But until recently, the overall strategy for social was still unclear. Furthermore, many organizations have continued to experience problems with their deployments—from weak use of taxonomy and templates to inconsistent governance and management standards. User adoption (people going to the platform) and engagement (using the platform to conduct their business activities) are ongoing management concerns, and the quality of collaboration on the platform is directly impacted by lack of adoption and engagement.

Cloud Enables Innovation

Thanks to the rapid adoption of cloud-based solutions, the way teams collaborate and connect inside and outside of the enterprise is changing. Tools that overtly manage and manipulate content are becoming more seamless and integrated across on-premises and online platforms. More than that, the lines between the tools we use at work and at home are blurring. Social is becoming less about a destination (going outside of your team site to a separate URL to have a social interaction with others) and more about delivering a seamless social experience that persists across workloads.

The maturity of Office 365 as a home for the next generation of social collaboration within the Microsoft universe is becoming better understood. The future of social, according to Microsoft, is all about the cloud. Even so, the social options outside of (or alongside) the Microsoft platform have also expanded, with several innovative companies offering an integrated social story across on-premises and online assets. These options allow customers to maintain their on-premises SharePoint platforms, getting value from existing investments, while also taking advantage of the latest social technologies to help drive adoption and innovation across the enterprise. Some organizations will turn to newer versions of the SharePoint platform in hopes that new features and capabilities will help fill the social gap. And other companies will consider these third-party solutions and tools as a way to engage end users and improve adoption.

According to AIIM.org president John Mancini, the world of content management has moved from a "system of record" to a "system of engagement" model. In this new model, social technology has become as important, if not more important, than the enterprise collaboration features that have driven the success of SharePoint and its competitors. He writes:

> "Forrester talks about how the combination of cloud, SaaS, mobile, social, and analytics dramatically changes the nature of collaboration, making it possible for the first time to truly address all of the grey areas of our business processes. They call this opportunity 'Smart Process Applications,' and I think there is a great deal to this line of thought. Every industry has processes that at first glance seem automated. And at the surface level they are. But the reality beneath the surface is that most processes have countless branches and outcroppings where right now lots of manual and ad hoc collaboration occurs. It is in the automation of these 'exceptions' that enormous opportunity lies."

As Mancini points out, our expectations about collaboration are changing as rapidly as the software and services that drive our enterprises. One clear benefit of social technologies is to provide a social fabric between our various business processes, allowing discussion — whether in real-time or through asynchronous communication — to bridge the gap between business workloads. The problem with collaboration, it seems, has little to do with the technology itself. It has more to do with the failure of organizations to align the technology with the business, and with the corporate culture.

The next version of SharePoint (SharePoint Server 2016 for on-premises environments, with Office 365 for the cloud) is not about the SharePoint we know and use today, but about the search and social services at its core. In a blog post earlier this year, Office Products GM Julia White spoke about the end-to-end "experiences" on which Microsoft is now focusing: the social interactions we have with our peers, our coworkers, and our customers across common workloads.

Social collaboration works best when it happens in the context of the work you are performing, and that is where Microsoft is focusing their efforts going forward. Many enterprises were on this path already, using free and consumer-based social platforms to fill the gaps between the structured collaboration platforms and their defined workloads. Microsoft has observed this shift in its own customer base, has learned from it, and is applying that learning to the next version of SharePoint and the Office 365 suite.

Think about the macro level of your collaborations, the increasing volume of content created. And even beyond the content, consider the data that is generated based on your social interactions, your search queries, your affiliations and transactional footprint. Microsoft is rapidly evolving their collaboration and social platforms, combining machine learning and cutting-edge business intelligence capabilities that draw from these various signals to provide a highly personalized experience for the end user. With the announcements for Office Graph and general availability of the Delve search interface, Microsoft is beginning to show signs of their future direction: data-driven, personalized, productivity-based solutions.

Besides search, the new Groups functionality is the first Microsoft foray into inline social experiences that enable social interactions within a single application, such as PowerPoint or OneNote. And Groups also lets you move with that same interaction across workloads—for example, a conversation started in Exchange might move across to Word or Excel, as needed, while remaining in context to the original interaction and in real time.

With the strategy change toward more social and search-centric solutions, observers expect to see additional changes to Microsoft product and engineering team processes. This may, in turn, increase the rate of change and innovation from the company. Specifically, Microsoft is investing more heavily in additional Data and Applied Science resources within each team to expand telemetry, improve processes and increase quality.

Speaking about this increased focus on data-driven prioritization, Microsoft CEO Satya Nadella commented, "I consider the job before us to be bolder and more ambitious than anything we have ever done."

Social Is at the Center of Productivity

At the 2014 Microsoft Worldwide Partner Conference, CEO Nadella shared a new vision for the company with a focus on platforms and productivity. For those of us within the SharePoint community, this change of focus was already well underway. With the release of SharePoint 2013, it was announced that all innovation would be focused on the cloud first, and that Office 365 would be at the forefront of the productivity strategy.

In a memorandum to employees in July 2014, Nadella stated:

"Productivity for us goes well beyond documents, spreadsheets and slides. We will reinvent productivity for people who are swimming in a growing sea of devices, apps, data and social networks. We will build the solutions that address the productivity needs of groups and entire organizations as well as individuals by putting them at the center of their computing experiences. We will shift the meaning of productivity beyond solely producing something to include empowering people with new insights. We will build tools to be more predictive, personal and helpful.

"Every experience Microsoft builds will understand the rich context of an individual at work and in life to help them organize and accomplish things with ease."[1]

Of course, technology is just one aspect of any productivity strategy. In the new world of information management and collaboration, where social plays such an important role in driving workplace productivity, organizations must consider how they will approach productivity as part of their corporate governance activities. As they seek to improve productivity within the enterprise, there are four key areas that every organization should clearly define and develop strategies for:

1. Information and Collaboration Management

Within the past decade, document collaboration has evolved from a team-based toolset to a business-critical requirement for the enterprise. As the leading document management and collaboration platform, Share-Point — and its online successor, Office 365, which also includes web-based email and communications services — has helped enterprises to break down many of the data silos surrounding their content and business processes. But while collaboration platforms are relatively easy to deploy and start using, most environments struggle with end user adoption and ongoing engagement due to a lack of alignment with key business processes, and a failure to understand the motivations of their end users.

2. Capture and Correlation

A knowledge management platform is only as good as the information it holds within. For example, a number of vendors have built their businesses on the capture of paper-based knowledge assets. And yet the volume of digital and rich media assets far surpasses the paper problem. In 2014, it was estimated that the average SharePoint farm contained over 1TB of data, but was growing 50 percent to 75 percent per year. As more and more organizations begin to focus on social interactions and Big Data assets, this rate of growth will only increase. It will become increasingly important to ensure all pertinent data is being captured and is identified in context to projects, customer data, business processes, legal and regulatory restrictors, and other related assets.

3. Social Interaction

With the rise of popularity of social networking platforms in the consumer space over the past decade, many enterprises began to look at these same social capabilities as a way to improve the "stickiness" of their knowledge management and collaboration platforms. As the capabilities of these tools matured, their use as communications tools has evolved as well, with most enterprise applications now including some sort of social capabilities. Instant messaging, once a stand-alone tool, has now become ubiquitous for internal communication. And SharePoint and competitive platforms provide inline social experiences that allow for contextual interactions within key workloads, such as when jointly editing a PowerPoint presentation.

The current reality is that these social technology tools, including Microsoft Yammer, Salesforce Chatter, and any other competitive social platform, generate massive amounts of valuable content. And they help our systems to better understand how we work and who we work with, providing yet another indispensable layer to our knowledge management platforms.

4. Search and Dissemination

Search has always been a central knowledge management concern. Storing content is relatively easy; creating a search platform that is easy to use and also powerful is both art and science. As the volume and complexity of our data has grown, the need for dynamic, powerful and personalized tools to help us search, find and share our content has become a business imperative. Microsoft understood the search requirement when it acquired

FAST Search in 2008. Microsoft built upon the technology for its 2010 and 2013 releases to develop a new family of search-based capabilities, such as Delve and Clutter, available through SharePoint Online (as part of Office 365).

Using machine learning and the data captured through social interactions, Delve offers a personalized view of an end user's content, organized by content currently accessed, by content shared by team members, or by learning from end-user activities. One of the problems with traditional search is that you need to know something about what you are looking for. Whereas Delve can surface data based on your interactions and the importance of that content to people within your close network. Clutter likewise uses machine learning to filter your email based on past history, reducing the amount of "noise" coming through your mailbox.

As you begin to think about the future of your own knowledge and collaboration management platforms, take the time for honest reflection about what tools and processes are helping you to identify, classify, contextualize and correlate your information assets. And identify which ones are hindering that process. The success of any project can hinge on having a shared understanding of what is to be achieved.

Extending Social Collaboration

A key to success in enterprise social has been to align social activities with specific business activities. An example of this is incorporating polling, threaded messaging, and ratings systems common in most enterprise social platforms into the product development processes, allowing the extended team to provide input into the identification and prioritization of features in a company's product roadmap. By extending discussion beyond the product development team (to include support, sales, marketing and possibly even customers), quality is improved (more specific, refined requirements). And expectations are better met (participants understand reasons behind architectural and feature decisions, and the timing of the next release).

While there are certainly ample scenarios for ad hoc, or unstructured collaboration activities (community building, idea creation), many organizations are recognizing the need for a more robust, structured collaboration model for their social activities. In other words, it's becoming not only more common, but expected, to have the ability to be "social" within the context of common enterprise workloads—such as records management, customer relationship management (CRM), or human resources-related activities. A quick threaded discussion could remove the need for hours of group planning sessions, reduce lengthy workflows, and help geographically dispersed teams stay in the loop.

One strategy for making collaboration analytics actionable in the organization is to introduce gamification techniques. In a nutshell, gamification is the process of measuring system behavior through community management, reward and loyalty programs, and game design. Gamification is, simply put, a set of tools to help you motivate and incentivize behavior on the system. Social-based gamification techniques are becoming more and more popular within corporate environments and public-facing sites as organizations look for ways to keep users engaged.

Whether you deploy simple chat-based social features or a feature-rich gamification platform, here are some strategies you should consider as you develop your own plan:

Collect the right data.
It's not simply a matter of gathering all data related to social interactions within your platform of choice. In the case of SharePoint, that could be a *massive* amount of data across numerous content databases. Understand the core metrics you want to measure, and the data you believe is needed, to accurately report on those metrics.

Identify the metrics that matter.
Identifying "the right metrics" can be a complex task in itself. Your metrics may be limited by the data that you have access to within your tools. But at the very least you can create assumptions and baselines for each metric, and refine them as you go.

An important aspect of tracking any metric is understanding how to take action on the data. A poor quality metric is one that includes no path forward, no way to make an improvement. Understand how to take action on what you learn (positive or negative), and refine your metrics as you learn more about how people are using the tools.

Review the data, and test your assumptions.
You've captured the data. You've created your baseline social collaboration metrics. And now it's time to share what you've learned — or, I should say, what you believe you have learned — with your management team and with your end users. Make the metrics transparent, share what you think the data means, and fold their feedback back into the testing and analysis process.

It could be that you need to "tweak" the data and your metrics because the results may not show you enough of what you want to know about your end users' behavior. Iteration is always a positive thing, because it means you're getting closer to the truth.

Give it some time, and watch the trends.
No metric at a single point in time is as valuable as the same metric shown over time. The value of the information is rarely in a snapshot, but rather in the upward or downward trends over time. When people know they're being measured, they tend to adjust their activities toward those metrics to make themselves look good — which is perfectly normal. So as people adjust to the metrics, you'll want to refine your analysis, and once again review your data sources and your metrics to ensure you're making the right assessments.

Fold what you learn into a strategy.
With data and metrics in hand, you're ready to develop an overall strategy for improving social collaboration within your organization. And by improving social collaboration, you enable your employees to be more productive, accomplish more work activity, and collaborate with purpose.

A sound strategy must first establish a set of proper expectations. Be open with your team and employees about what you are trying to achieve, and work closely with them to develop the right solutions that will motivate the right behaviors. Your strategy must focus on your end-user objectives — the desired behavior — and align with business processes. And it is also critical to align incentives and rewards with team and corporate culture, ensuring they will truly motivate your team. For example, many sales reps are driven by leaderboards (another gamification technique) with highly competitive rewards by day, week, month, quarter and year. But that strategy may not span other parts of the company (marketing, operations, engineering) if your goal is to create programs across the entire company.

The key to success for managing all of these different tools and platforms can best be summarized by these five tips for IT administrators and business users. Keep these tips in mind when collaborating across multiple platforms:

Define how each tool is to be used.
Understand the primary use cases for each tool, and their target users. Each solution has specific strengths and clearly defined use cases. Chatter, for example, might be the right place for sales teams to communicate; but it is probably not the best platform to use as your system of record for storing content. A huge benefit of using Dynamics CRM over Salesforce is your ability to embed Yammer conversations within the platform. Those same conversations (Yammer Groups) can then be embedded within SharePoint, making all customer conversations in Dynamics visible across SharePoint and the Office 365 platform.

Talk to your end users about these use cases. If their expectations extend beyond the proper use of the platforms, use this as an opportunity to educate them on other available tools or document these unmet requirements as you investigate additional solutions.

Understand the boundaries of each platform.
Do you know the storage limits for each list and library? Does the tool require performance maintenance to ensure quality of service? Are there reports, metrics, or flexible configuration settings that enable you to better manage these tools? Know the limits of each tool, take the time to do some capacity planning for their expected growth, and figure out how you will support your end users going forward.

Manage (encourage) engagement.
You've worked with your end users to understand how they plan to use the tools. You've done your homework on how users interact and what support is needed to keep them engaged. But suddenly, usage drops. What went wrong?

Ensure visibility measures are in place to monitor how your employees are using their collaboration tools, and talk to them about what is working and what is not. In the early stages of any new technology deployment, there will be adjustments as you and your end users adapt to them. Encourage feedback and stay ahead of the change management process.

Monitor adoption.
Keep track of overall activity to get a sense of user adoption. Some tools have a short-term life as the needs of the business change and as end-user workloads shift. This is the reality (and the downside) of the consumerization of IT: fickle end users. It is highly recommend that companies pilot before full-scale deployments of new collaboration platforms.

Lead, don't follow.
This one is less clear and more forward-looking, but still important. Watch for end-user behaviors that may lead to adoption of new technologies. Widespread use of Facebook is a good indicator that your end users may want or need some kind of social collaboration platform, but in a slightly more secure manner. Instead of being reactive to these changing trends, keep abreast of new technologies. Work with your end users to try out new tools and platforms, experimenting with ways to improve overall collaboration. If you build a culture of experimentation and trust, your end users are more willing to provide feedback on their collaboration habits and usage.

The rapid increase in collaborative technologies is exciting. We are witnessing a dramatic change in the way that information workers access and relate to content, and how they interact through team collaboration environments. Take advantage of these new tools and platforms. Benefit from their ability to spark innovation across your business to drive engagement, improve collaboration and instill a stronger sense of community.

It's easy to get carried away with metrics and gamification techniques to the point where you lose sight of what really matters to your business. Listening to your end users is important. But essential to the long-term health and well-being of your business interests is to have a more balanced approach using sound governance principles and flexible solutions for your employees. You want to encourage employee collaboration and enhance their productivity, but at the same time mitigate risks across the systems they use. The goal should be to stay in tune with changing trends and new technology through visibility and control processes. Experiment with new initiatives that help you innovate and keep your end users engaged — always with the focus on driving value to the business.

NOTE
[1] *Read more at http://bit.ly/1qNqtP3*

ABOUT THE AUTHOR

Christian Buckley is an Office 365 MVP and Managing Director, Americas for GTconsult, a consulting and managed services provider with offices in the US and South Africa that specializes in "Everything SharePoint." Over the last several years, he was instrumental in the acquisitions of two SharePoint ISVs (echoTechnology in 2010, and Axceler in 2013) and helped build some of the most recognized product brands as Chief Evangelist at Axceler and Metalogix. He previously worked at Microsoft as part of the enterprise hosted SharePoint platform team (now part of Office365), and led an engineering team in advertising operations. He can be reached at cbuck@gtconsult.com and www.twitter.com/buckleyplanet.

Chapter 10

Applying Enterprise Social Graphing to Transform Employee and Customer Experiences

by Naomi Moneypenny

Introduction

This is an extraordinarily exciting time for enterprise software. A true revolution is occurring. It was initially heavily influenced by the explosive adoption of social networking in the consumer world in the past decade, led by companies such as Facebook, Twitter, and LinkedIn. But this revolution is now on its own trajectory—a trajectory that is thankfully optimized for the particular requirements of organizations, not just individuals.

This revolution rests on the seemingly humble concept of the *graph*. The term *graph* is derived from the mathematical field of graph theory, which is focused on the structures and behaviors of various types of *networks*. This makes perfect sense, because the revolution in enterprise software is in many ways very much about how the basic system architectural elements transform from hierarchies into networks.

You have probably heard of the *social graph*, which is simply a fancy term for all of the connections among people in a social network. It was a term originally introduced by Facebook, but it applies to any system in which people connect with one another. Those connections can be symmetric (e.g., *friending*) or asymmetric (e.g., *following*). The individual people in social networks can be represented as nodes in a graph. The connections are represented in graph theory as *edges*. These connections or edges can have additional attributes, such as distinct directions (e.g., following) and even weightings. Encoded in this convenient form, the graph can provide various insights. For example, you can identify common connections among two or more people, as well as the well-known degree of separation between any two people. More sophisticated network characteristics can be derived from the social graph, but even these simple constructs can be quite usefully applied. For example, a system can suggest that you connect with someone who is connected to one or more people in common with you. This is the basic approach behind the people suggestions you receive on Facebook, LinkedIn and Twitter. It is also commonly used in many enterprise social platforms, such as SharePoint, Yammer and Office 365. More complex approaches can combine classic search, which matches a search term with the content of documents, with social graph information. Facebook's Graph Search is an example of this approach. The more recent versions of SharePoint Search also embrace this technique.

The power of social networking and the social graph has taken the enterprise by storm. The trend accelerated after Microsoft acquired Yammer. This move made it clear that social networks were going to be much more than just another enterprise application — they were destined to become a fundamental and ubiquitous capability.

The Social Graph Extended

But the social graph is just a subset of the more general notion of a graph of relationships among *computer represented-objects* in general. The objects can represent people, as in classic social networks, or content, such as documents. And it is this more general type of graph that delivers maximum value to organizations. This is because it effectively integrates social networking and content management into a cohesive and dynamic network of knowledge and expertise, which can propel organizational effectiveness to new levels.

So this is the first important extension to the social graph that is necessary to deliver greatest possible value to the enterprise: generalizing the graph to include people and other computer-represented objects, such as content. The second required extension involves automating edge and connection data. Originally the edges of the social graph were limited to explicitly determined relationships, such as following or friending actions. However, edges can also be determined automatically. Automatically determined edges can be of various types, have particular directions, and have weightings. These automatically determined edges can be based on usage behaviors or inferences thereof. Even before Facebook popularized the social graph, my colleague Steve Flinn and I invented[1] this more general approach that is particularly suited for enterprise applications. In a network of people and content, the connections and edges among and between them are automatically and continuously adjusted based on usage behaviors.

In such integrated networks, relationships between people and objects (other people or items of content) can be represented in the form of Actor-Edge-Object *triples*. The Edge can denote an action or a type of relationship. The Object can represent another person or an item of content. So, for example, a user, Jane, viewing a document, could be represented as the behavioral-based triple, Jane-Viewed-Document1. Or in another example, Jane might follow Tom to receive Tom's social networking posts. This could be encoded as Jane-Follows-Tom. An enterprise collaborative or social networking platform generates vast numbers of these types of Actor-Edge-Object triples (along with auxiliary information such as timestamps). And such triples constitute the social big data-based raw material on which anticipatory computing rests. In these two examples of Actor-Edge-Object triples, the Edge represents actions or relationships that are explicitly performed by a person: viewing and following. But edges can also encode algorithmically determined values associated with relationships between people and other people or objects.

An example of this this type of integrated social and content network in the consumer space is Facebook's Open Graph. A more recent example of this this type of integrated social and content network in the enterprise is Microsoft Office Graph, available in Office 365. Office Graph is fundamentally an integrated network of objects (which can represent both people, or *actors*, and content). Some of these objects are weighted, and these weights are continuously adjusted based on users' interactions with the various objects. As depicted in Figure 1, the Actor-Edge-Object triple is the fundamental structural unit of Office Graph. Office Graph can be accessed by an API (the Graph Query Language) to provide the data necessary to make automatic inferences and generate other useful graphs, such as an interest graph. It is clear that Office Graph will be fundamental to Microsoft cloud-based enterprise software directions.

Figure 1. Office Graph Actor-Edge-Object Triple

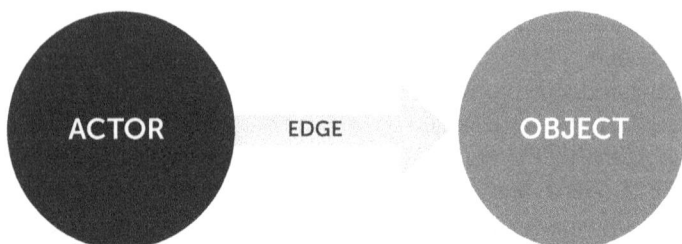

Another example is a product from my company, ManyWorlds. Our Synxi-brand anticipatory computing-based apps and connector products integrate enterprise collaborative and social networking platforms, including SharePoint and Yammer. As with Office Graph, the Synxi architecture is underpinned by weighted integrated social and content network structures that are continuously updated based on user activities. Synxi provides an API that can be accessed to surface aspects of the graph for various purposes, but with distinguishing features with respect to the environments for which it applies and the functional capabilities provided.

Graph Analytics and Metrics

Integrated social and content graphs engender a wide range of new and valuable ways of measuring objectives that are important to us. For example, you can easily extract individual and community adoption metrics from the graph. These metrics can gauge the overall amount of interactions occurring within a community, which participants have the greatest quantity of content contributions, or who interacts most with published content. You can also easily determine popularity factors associated with particular content (as measured by views and/or other types of interactions with the content).

For example, one of the balancing acts facing all organizations is the trade-off between *structure* and *serendipity*. Structure provides the rules-of-the-road that ensure the efficiency and consistency required for sustaining high organizational productivity. Serendipity is all about the unintended connections that occur outside our normal patterns that foster new ideas and help drive improvements and growth. If there is too much structure, serendipity can suffer—the structural pavement of our daily cow paths inhibits trying out new things that can lead to improvements. On the other hand, insufficient structure is the path to inefficiency and chaos. The notion of finding the right point in this trade-off when managing organizations is an example of "loose-tight" management—managing both loosely enough to enable serendipity and its resulting innovations, while managing tightly enough to ensure a smoothly operating organization.

So as social platform adoption progresses, what measurements from the underlying graph enable us to gauge whether a good balance has been struck between structure and serendipity? Well, measures of associated structure should include the appropriate organizing of people and content with groups and topics, or their equivalent. In particular, *nearly all* content items should be associated with *at least one* group or topic, and preferably more than one. There should also be a reasonable *distribution* of content among groups and topics. Groups or topics that are associated with nearly everything or that are associated with nearly nothing are not very useful. So some good measurements of adequate structure for mature social deployments include:

- **Orphan posts:** Posts *not* associated with a group or topic
- **Nurtured posts:** Posts associated with *more than one* group or topic
- **Structural richness:** The ratio of the total number of groups and topics to the total number of posts

These metrics should be at least a good starting point in ensuring that you have a structure that supports a high performing organization. Of course, more sophisticated measurements of the dispersion of content to groups and topics are also possible.

Whereas structure metrics can essentially be *snapshots* in time, measuring serendipity, in a sense, requires a *video* of activities. You want to get a handle on the dynamics of the *growth* in connections among people and content and *flows* of information propagating along the graph. So the kinds of behavioral information you have at your disposal to gauge serendipity include sub-community membership and following patterns, as well as interactions with posts. Good measurements to gauge adequate serendipity for mature deployments leveraging these data include:

- **Joining and following growth:** The ratio of the number of *new* group joins and follows in the last 90 days to *total* group joins and follows.
- **Engagement breadth:** The percentage of posts that receive at least one reply or like during the last 90 days
- **Engagement depth:** The ratio of total likes and comments to total posts during the last 90 days

By publicizing and applying these metrics, you can tune the mature social deployment to simultaneously promote efficiency and vibrancy. For example, if there are too many orphan posts and not enough nurtured posts, you may want to remind users of their responsibility to appropriately categorize their content with groups and/or topics for the benefit of the rest of the community. If following growth has stagnated, or if engagement breadth or depth is lagging, you can remind users that it is sometimes beneficial to prune their networks by unfollowing, when appropriate. As discussed below, discovery tools that automatically surface relevant posts can be helpful if personal signal/noise ratios begin to fall.

Influence and Expertise in Social Graphing

Because of the rich interconnections between objects in the graph, we can actually go even further in generating valuable metrics. We can also make cumulative calculations along chains of edges of the graph to generate more sophisticated and often more meaningful metrics. An example of this is an *influence* metric. We all have an intuitive sense of what influence is: the capacity to affect the behavior of other people, obviously a very powerful characteristic to possess. Those who have it are therefore very valuable to those whose business it is to influence others. Because if the influencer is influenced, many others will be influenced as well. So influencers are often targeted by marketers to take advantage of this amplification effect—it gives them the most bang for the marketing buck. Marketers who covet influencers are not necessarily just those who have a product or service to sell to a consumer. We also "market" objectives within an organization, such as helping to promote the adoption of a new process or technology.

Traditionally, influence could be somewhat slippery to actually measure. But with our integrated social and content graph, we can quite effectively quantify influence. Such quantifications can be made with varying degrees of sophistication. *Reach* is one of the simpler such metrics—for example, the number of followers one has. A more sophisticated, *recursive* approach, which iterates across graph edges, not only takes into account the number of followers one has, but also the *influence of those followers*. (And their influence is likewise a function of the influence of their followers, and so on.) Other behaviors[2] in addition to following can be considered in quantifying influence as well. Examples include the degree to which actions by others are taken with respect to content published by a person, and actions such as comments, likes, re-sharing, etc. Influence can be continuously calculated in such a manner and displayed in real time. This can provide useful insights into the go-to people to get on board for an initiative, for instance. When disseminated widely throughout a network of social network users, influence metrics naturally encourage people to increase their influence. This in turn drives the type of collaborative behaviors that benefit the entire social community.

Understanding and measuring influence is important in the consumer world. But with the advent of graph-based structures in the enterprise, it is becoming equally important in the contemporary workplace as well. There is another personal quality that is also particularly important in the enterprise: expertise. Expertise is the quality of having a skill or competence in an area. Expertise and influence can overlap, but not necessarily. One can be expert in an area, but not be very influential, and vice versa.

As with influence, expertise—particularly relative expertise—can be inferred from social behaviors. In fact, expertise can be assessed from some of the same kinds of behaviors as influence is, but viewed from a different perspective. For example, someone with many followers tells us something about their influence, but it does not tell us much about their expertise on any given topic. On the other hand, someone having followers with above average expertise on a topic may very well tell us something about the expertise of the person being followed. As another example, publishing content that is rated highly by others—particularly by others who are already determined to have higher-than-average expertise—can also be indicative of expertise.

Of course, expertise can also be assessed directly by scraping information from people's profiles. And that may at times be a reasonable starting point in assessing expertise. But ultimately, actions speak louder than words. Expertise inferences derived from actual behaviors within the graph structure enable finer-grained, more credible, and up-to-date assessments of relative expertise than merely using profile information. Such a behavioral-based approach enables you to identify those with relevant expertise in real time, and at a granular level of relevancy that enables expertise matching with topics associated with a specific document.

Anticipatory Computing

The integrated social and content graph enables a whole set of valuable analytics and metrics that are useful for people. But these analytics enable much more. The integrated graph and associated analytics make possible a highly beneficial system capability: continuously *anticipating* what users need, rather than merely reacting to users. This ability to automatically anticipate what users will find interesting, based on what a system has learned about them over time, has become a core feature of most major consumer platforms. Google, Facebook, Amazon, LinkedIn, and Netflix are just some of the most prominent of such platforms. These companies all capture significant volumes of collective and individual behavioral information, and then apply machine-learning-based algorithms to make unprompted recommendations of products, content, other people, and groups.

This machine-learning-based *anticipatory computing* capability is more than just a useful technical feature — it is now fundamental to the *business models* of these companies. For example, Amazon reportedly derives over 30 percent of their revenue from their recommendations. LinkedIn reports that over 50 percent of the connections made between people and between people and jobs are via their automatically-generated recommendations. Netflix drives 75 percent of their movie rentals from their recommendations. It would be unfathomable for a major consumer platform today to not be applying automatic learning engines to deliver unprompted, personally relevant suggestions derived from the social big data that is available.

So what about the enterprise? Well, historically the behavioral information on which anticipatory computing rests has simply not been captured and available in the enterprise. And without that information personalized, anticipatory computing is impossible. Unfortunately, the sheer quantities of content and the complexity of relationships in most organizations inevitably create tremendous friction and inefficiencies in ensuring the right knowledge and expertise is available to the right person at the right time. The symptoms of this complexity crisis are visible in user exasperation feedback such as, "poor findability," "information fire hose," "poor signal/noise ratio," etc. Think of the Netflix example: people in the enterprise may be missing out on up to 75 percent of what they need, while getting a lot of what they don't need! Fortunately the situation has changed radically for the better with the availability of more advanced collaborative platforms and enterprise social networks (ESNs). These newer systems have the graph structures and the ability to capture the information required for anticipatory computing.

How Anticipatory Computing Works

The integrated social and content graph forms the structure for anticipatory computing. The technique works by generating and continuously adjusting values associated with the connections or edges of the graph, based on social big data-based analytics. These continuous adjustments constitute *learning*, just as adjusting connections of neurons in our brain from the external information that we humans process constitutes learning. As depicted in Figure 2, what a human or a system knows at any given time is static. Learning, on the other hand, whether human- or machine-based, is a function of the change in the states of knowledge over time.

Figure 2. Learning Embodied in Changes to Network Connections

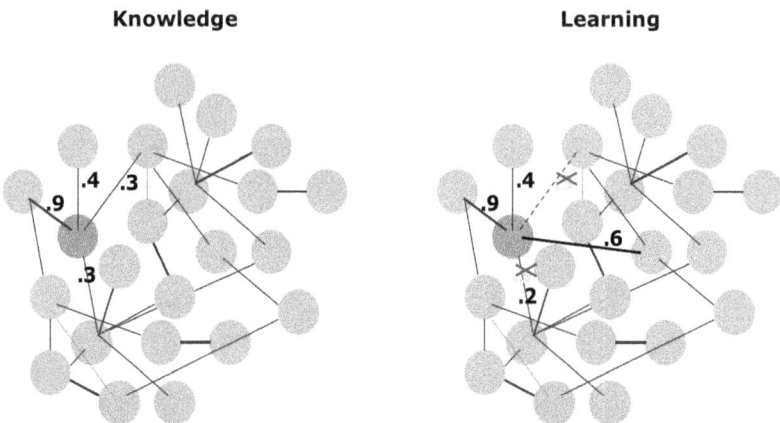

Knowledge

Learning

This type of machine-based learning can be applied to generate auxiliary types of graphs that form the foundation for anticipatory computing. For example, you can generate an *interest graph* for each user. An interest graph is a set of interest-based relationships between a person and something else, typically a topical area. These interests can be explicitly indicated by the person. But interests are more often inferred from other Actor-Edge-Object data that encode various user actions, such as postings, views, likes, commenting, and following. Whether explicitly indicated or implicitly inferred, interests are generally a matter of degree. So interest relationships are typically represented as a continuum (usually normalized to a 0-1 range) rather than as a black or white, yes or no, interest relationship.

The interest graph, often in combination with the basic social graph, enables a whole array of new and very useful functionality. For example, an item of content can be suggested to you based on the specific degree of interest that is inferred from topics associated with the item of content. Or another person can be suggested to you, not only based on your social graph, but by also taking into account an inferred similarity of interests between you and that other person. Most fundamentally, the interest graph enables a capability for automatic personalization, which is the core of where *all* user-oriented computing is headed, whether consumer or business-related.

Besides interest, at least one other relationship between people and topical areas can be valuably applied for anticipatory computing purposes: *expertise*. As already discussed, a person's expertise level on a topic can be explicitly indicated by the person or by other people (e.g., endorsements). Or it can be implicitly inferred from the person's actions, from other people's interactions with the person, or from the person's works. Similar to interests, levels of expertise on a topic are generally a matter of degree, and are therefore best represented as a continuum. The resulting *expertise graph* can be put to work on tasks. An example would be automatically matching in real time the people who are predicted to have expertise in one or more areas with people, projects or jobs that require expertise in those areas.

Just as the social graph is continuously updated as people add or modify connections, interest and expertise graphs are continuously updated by an anticipatory computing system based on inferences from people's actions. These updates are performed by machine-learning-based algorithms that operate on the social big data generated by associated social networking and collaborative systems. Taken together, the social, interest, and expertise graphs lie at the very heart of the systems that we can expect to be working with going forward. In the consumer world or in the enterprise, these systems automatically and continuously learn from us, thereby anticipating and delivering to us the right knowledge and expertise we need right when we need it.

Anticipatory Computing in Practice

The integrated social and content graph, along with machine-learning-derived interest and expertise graphs, provide the foundation for anticipatory computing. But an anticipatory computing user interface that directly provides users personalized discovery is also required. For Office Graph, this user interface is the Delve application. For Synxi, personalized discovery can be integrated directly into the SharePoint user interface. Or it can be delivered in a separate universal interface for personalized discovery that can span Yammer, SharePoint and other ESNs. Synxi can generate recommendations that are not only personalized, but that are also relevant to a particular user context (e.g., viewing a specific document). Synxi also constructs detailed expertise graphs that are continuously adjusted based on what people actually *do* rather than merely relying on explicit profiling information, which inevitably suffers from the drawbacks of incompleteness and/or lack of freshness, not to mention credibility. Such Synxi expertise graphs can then be used to recommend other users whose automatically generated expertise profiles match the subjects associated with a particular document. Synxi expertise scores can also be visually displayed in various ways, as shown in Figure 3.

Figure 3. Contextually Relevant Suggestions Based on Automatically Inferred Expertise Levels (Data from Microsoft Office 365 Yammer Network)

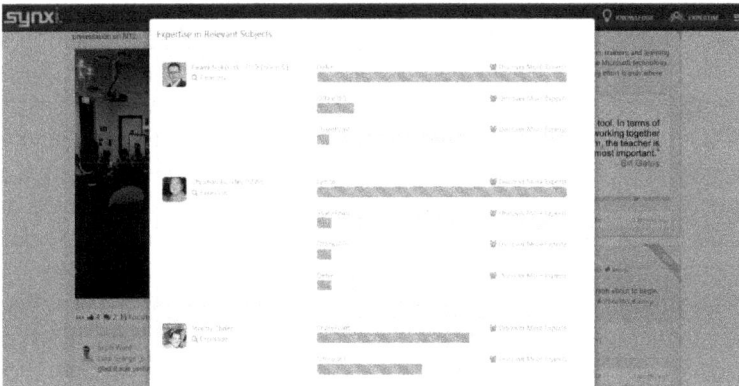

Beyond Your Enterprise

We've looked at how social graphing is an important data source inside your company, assisting in use cases such as expertise discovery, talent management and contextualized recommendations. Now let's go beyond your own company to think about how this data is useful in deriving insight about customers, products and other potential investment areas from a growth perspective.

The first step when expanding your social graphing approach is to use a feature such as External Networks in Yammer. Say you are in a services business, where partnering with key customers is critical. It is very important to understand the needs of these key customers, perhaps even the trending in their concerns, interests and potential directions. Extending the graphing approach to these key relationships is a source of insight as well as a way to understand how well your partnership is working.

We can take this idea further and think about a consumer approach. Here are a couple of examples that I often talk about in my presentations. They help illustrate the point of what you can do with data.

Let's begin with Amazon.com, one of the pioneers of collaborative filtering. The reviews on Amazon are probably the most important element of customer data that Amazon curates. And that information is useful to us regardless of whether we choose to transact on Amazon or not. But it is not just you who reads the reviews on Amazon to inform your personal buying habits. That free information is a source of revenue to others. In the last few years, small companies have sprung up to mine that data and create a new system of product design and fulfillment that happens at such speed that traditionally structured companies cannot keep up.

For example, a small 150-person company in New Jersey runs a product development, design and fulfillment process entirely through Amazon. By reading the reviews on Amazon, product managers mine insights on what customers want—a wireless speaker for example. They read that customers want the speaker to be available in a choice of colors, use Bluetooth, be waterproof to use near a pool or in a shower, or have a rechargeable battery. The product managers collect the most requested features and build a specification sheet. Those product managers then send the specifications to be built in China or wherever the lowest cost per feature can be manufactured, and the items are drop shipped to Amazon's warehouses. The product managers then list the items on Amazon for sale, which are surfaced by customers searching for "wireless speaker." And the entire purchase, fulfillment, and even any potential returns are handled by Amazon. As of now, this isn't an automated process; but you can clearly see how easy it is to gain insights into what people want. This New Jersey-based company has revenues of nine digits a year, and at last report was growing at 30 percent a year.

Let's take another example. Twitter released some data on holiday shopping. Mining the data on Twitter may seem far-fetched to some, but in this new world of building products that people want, the ability to segment, slice and dice, and detect profitable niches is critical. And it's already big business, with more than 190 million visitors to Twitter's site every month. In a study of 2000 U.S. shoppers, more than half used Twitter regularly. Thirty percent of those users began thinking and tweeting about holiday shopping before October. Two thirds of the shoppers discovered new products via Twitter and importantly, 64 percent of them purchased a product because of Twitter. And then for extra credit, 62 percent of the users tweet about the purchase that they made, both for feedback and viral stimulation. You can just imagine how creating the holiday must-have toy, accessory or service trend is possible just by mining the data from Twitter.

But the most bang for your buck comes from combining what you know inside your company with these sources of open data. Regardless of whether you are in a business-to-business situation or business-to-consumer, blending insights from the social graphs of your customers with those of your employees is a profitable path to sustainable growth. Inditex, the parent company of fashion retailer Zara, is one of the most innovative retailers in the world and has a market cap of over $94 billion. Every day managers in Zara's 6000 stores worldwide chat with customers to gather their feedback on the items they sell. They collect this feedback in their enterprise social network and this data is also reconciled with the point-of-sale data. The product life cycle is dramatically shortened with a team of only 200 designers for the entire chain, who produce more than 11,000 designs per year. That's at least 3 or 4 times more than competitors. The time to produce garments in a vertically integrated chain is accelerated, and with customers as part of the design process, the typical journey from runway to retail takes only 4 weeks. In Spain, the home country of Zara, the average retail store expects to see a customer three times a year. For Zara that metric is an average of 17 visits a year. And in combination with social media technologies like Facebook and Twitter, consumers sharing purchases which they may have helped to design and create typifies a trend that only serves to stimulate further demand.

We can see the far-reaching implications of social graphing. Inside your enterprise, it begins as a way to listen to explicit and implicit signals from your workforce, to help them be more productive as well anticipate the resources they need. Then it can blossom to your customers, suppliers and partners, and even to the end consumers, who might indirectly affect what your business does and how it does it.

What's Ahead?

As social and collaborative systems become more prevalent in organizations, they become more intertwined with other systems, ultimately becoming not just independent applications, but an enterprise IT architectural *layer*. Emerging from this base social layer is an even newer layer of integrated social and content graphs. When combined with machine learning, a new enterprise IT architectural layer emerges: an integrated machine-learning-based layer. This new layer learns from its experiences with users across applications by taking meaning on a continuous basis from the social big data that is generated. This integrated graph and machine-learning-based anticipatory computing layer can simply be termed *The Learning Layer*. This is also the title of the recent book that details the inevitable advances of graph-based structures and anticipatory computing in organizations, and explores the strategic impact these capabilities will have. As shown in Figure 4, the emergence of the social and leaning layers as bona fide enterprise IT architectural layers heralds the era of a truly personally adaptive IT stack. That personal adaptation extends to all the systems we interact with, whether we touch them through our work computers or the devices we wear to monitor our lifestyle. The old cliché "relationships are everything" takes on a new meaning when we extend that to the graph of data we produce and consume and the devices in the Internet of Things that help us accomplish tasks with ever-increasing efficiency.

Figure 4. Enterprise IT Social and Learning Architectural Layers

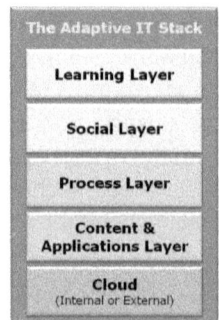

NOTES

[1] *Flinn, Steven, and Naomi Moneypenny. "Adaptive Recombinant Systems." World Intellectual Property Organization. Publication no. WO/2005/054982, June 16, 2005.*

[2] *See U.S. Patent No. 8,600,920, "Affinity Propagation in Adaptive Network-based Systems" for variations of such recursive approaches to calculating influence that we have invented.*

ABOUT THE AUTHOR

Naomi Moneypenny is Chief Technology Officer at Synxi (a Many-Worlds brand) where she leads the research and development team for machine learning-based personalized knowledge and expertise discovery for enterprise collaboration systems. She has been awarded over 30 patents in the field of adaptive systems and anticipatory computing.

Naomi is a Microsoft Office 365 MVP and was named a Top 25 SharePoint Influencer in 2014. She has led global deployments of Enterprise Social networks and advises some of the world's largest companies on how to become more agile and innovative through the application of new technologies.

Prior to ManyWorlds, Naomi led the technology forecasting process at Royal Dutch Shell and worked on the knowledge management framework that was applied globally to the company. An astrophysicist by background, Naomi is a frequently requested speaker at international events for her expertise in enterprise social & collaboration, anticipatory computing & business growth from innovative technology. Follow Naomi at @nmoneypenny and read her latest thinking at NaomiMoneypenny.com.

https://StreamingItOutLoud.com is a great tool to drive adoption and engagement of your Yammer network—used by over 300 organizations.

Learn about Expertise and Knowledge Discovery for Yammer, SharePoint, Office 365 and beyond at http://Synxi.com.

Chapter 11

Using Analytics to Drive Technology Decisions and Deliver Dynamic User Experiences

by Kunaal Kapoor and Antoinette Houston

Introduction

Analytics have traditionally been used to inform marketing and strategy decisions. Their importance in usability and user experience has increased significantly as organizations look to become more diverse in the way they access enterprise information through the use of different devices, channels and mediums.

The ability to incorporate quantitative and qualitative data from analytics in an effort to aid research and design is not only invaluable, but instrumental when trying to develop exceptional user experiences. Given the fact users are often spread across the globe, using a diverse set of analytics helps to bridge knowledge gaps and fully gauge and understand unique problems. Recently, Gartner, Inc.[1] identified *analytics* as one of the key usage trends of this year (2015) for enterprises, clearly outlining the importance of understanding user behavior data in helping to deliver **the right information to the right person, at the right time.**

BrightStarr has delivered award-winning[2] intranets, and as such, understands some of the common challenges applicable to most enterprises. Some of the top few trending challenges within organizations include:

- Understanding the needs of their employees
- Positively engaging and motivating their users
- Focusing on their user's top tasks
- Monitoring the performance of their systems and solutions

With increased globalization of organizations, employee expectations from digital collaboration platforms like SharePoint are becoming more of a challenge to predict. As a result, the use of analytics is one of the primary ways to represent key facts in an effort to resolve these common enterprise challenges. Moreover, a clear win occurs once organizations use this data to help clients design roadmaps to maximize their return on investment and ensure their systems grow and evolve with their business.

In this chapter we will explain how our consultants use analytics in two core ways:

- To **prove** why an upgrade is needed from an existing intranet, and
- To **improve** an existing intranet in order to increase adoption and usage.

We validate the need for analytics by assisting our clients in understanding the needs of their end users, but also how to cater to those needs. Using relevant analytics, our clients can better visualize and inform effective plans aimed at achieving enhanced productivity throughout their organization. We have provided several examples of how we have used analytics to drive employee engagement and achieve the ROI our clients seek from their investments in technology updates or in improving the user experience via enhancements to existing features on the intranet.

Our Approach to Drive Technology Decisions and Deliver Great User Experiences

The seeds of a successful project are sown at the very beginning. Understanding the needs of an organization in order to make informed design and technology decisions is critical. In fact, in our own work, getting to know our client is a process we place the utmost importance on. Our experts focus on this from day one of an engagement. We believe a company's success is founded primarily on its people, not just the technology they use. By analyzing the needs of employees at all levels of the business, and analyzing the processes and interactions occurring between them, our consultants are able to successfully engage our clients and understand their user goals and organizational goals. We put so much emphasis on this needs analysis for our clients because we know it is critical. Think about whether you are putting enough importance on getting to know your own internal stakeholders and their needs.

Understanding the Present and Planning for the Future

In the world of User Experience (UX), analysts are constantly reviewing, analyzing, and optimizing multi-channel experiences. The more educated they are about their user landscape and the appropriate tools needed to recognize and determine their user's expectations, the better the experience. Whether they use personas to help gauge and recognize user behavior, or analytics to help determine user environments and usage baselines, the breadth of analytics provides UX practitioners a variety of toolsets to prove and improve user experiences.

First, try to understand the current status of your client's business and their operations. Ask questions such as:

* What processes are in place?
* What systems are currently being used?
* How do employees currently work?

Only once the current situation is understood and a clear picture has been painted, will they move on to planning for the future. We believe understanding the "here and now" is the key to building better systems and implementing smarter processes. Employee engagement for any enterprise varies based on several factors. However, one common factor that aids in creating the right strategy and ensuring success is making sure users can complete desirable actions quickly and easily.

Understanding Your Target Audience: Personas

When working with organizations that have employees in global locations, it becomes very difficult to have representation for all users in an intranet upgrade effort. We believe if you don't have a clear understanding of user goals, user needs, current pain points and desired content, you will not see a high adoption rate of the new solution. Our resolution to this issue is the use of *personas*.

A persona is not a real individual, but rather a synthesized fictional character representative of a unique group of users who share common goals. We base our personas on qualitative, ethnographic user research, combined with other environmental and demographic details to include accurate behaviors, attitudes and needs of real users.

There are many industry variants concerning the viability of personas. But we believe without them, organizations risk creating a design not well suited for their target audience. When organizations lack ample knowledge of the various characteristics of the types of user groups visiting their sites and using their systems, it becomes harder to design an experience that includes all the key elements each type of user needs. The result is often a solution which fails to properly meet user expectations.

Figure 1. Identifying a Persona

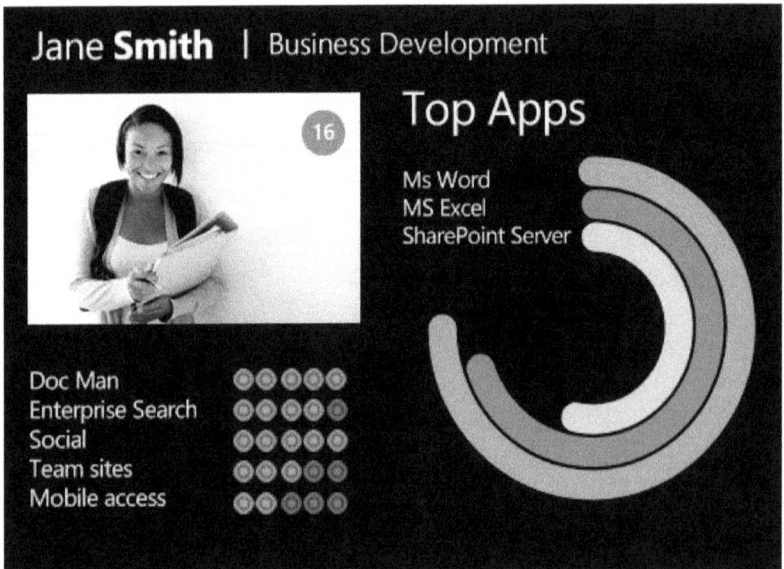

We consider personas a useful way of keeping users at the center of every-thing we do. And experience has taught us that user-centric designs are key to successful applications. Using specific information about the target audience of the applications, we are able to have focused discussions with stakeholders about prioritization and ROI, and build an intranet that meets expectations.

It's too easy to get tangled up in the complexities of new technology and lose sight of the goal: to make employees more engaged, more motivated, and more efficient. Focusing on these targets rather than competing for attention against the "next big thing" (social, apps, the cloud, etc.) helps increase and maintain user adoption, in addition to matching general expectations of having an intranet available for users. This in turn fosters increased collaboration, and facilitates users being more efficient and effective at work.

Segmenting Analytics Data Using Personas

Using analytics combined with personas, you can analyze how real users utilize the intranet. By creating persona-based segments in your analytics, you can determine differences in behavior between users groups. You can also understand the behavior of users who visit the intranet during peak times (e.g., normal office hours versus evenings and weekends). Moreover, you can attain details on user pain points that may be location-specific, and gauge interest in specific content for specific roles within the organization.

Using persona-based segments of analytical data, you can create behav-ior patterns. You can then use these behavior patterns to provide targeted information, optimize content based on device types, and even set prefer-ences for specific operations that users may need to perform on the site.

Segmented data can be used to analyze how users employ existing fea-tures, and to help determine which additional features or modifications might enhance the user's overall intranet experience. This kind of data can also help set priorities on core tasks and provide additional support to top-level tasks throughout the solution.

Prove It – ROI of Upgrades and Enhancements

We believe a commitment to the Microsoft enterprise stack is a long-term investment in both technology and the knowledge of how best to use it. Our approach involves:

- Helping organizations develop strategic roadmaps
- Highlighting why they need to upgrade to newer technologies
- Mapping the features and functionality that can be easily delivered to them using the latest tools available
- Mapping technology trends, user needs and evolution strategies using analytical data, segmented by personas, that represent target users of the application

Using this information we maximize the return on investment for our clients and ensure their systems grow and evolve with their business.

The following are some examples of how we justify the need for new technology platforms, as well as how we improve certain features using analytics:

Improving Search & Findability — The Need to Add FAST Search and Search Enhancements

After studying a client's analytics report on search and findability of content for end-users, we realized the organization (Figure 2) needed a more robust and intuitive search experience. There was a need to make the task of finding frequently searched items easier for line-of-business applications. In addition, facility-specific information must be easier for new employees to find. Using the information gathered from the most frequent search patterns, we recommended adding some new features to improve the user experience.

Given that this organization was leveraging the SharePoint 2010 platform, it was recommended they upgrade to FAST Search to take advantage of its benefits. We also recommended and implemented additional functionality, such as the ability to provide each employee with a list of easily accessible and personalized quick links to store frequently used resources. Another successful recommendation was the inclusion of a search-term tag cloud populated with frequently searched items, allowing users to click and search without typing in the search box.

Figure 2. Search-Term Tag Cloud

Search

Search Phrases Found

ADP ENGINEERING%20ESSENTIALS
FACILITIES HOLIDAY%20SCHEDULE
JOB%20OPENINGS **LEARNING%20CONNECT**
PAYROLL PDP
SALES%20SOURCE W2

Search Phrases Not Found

EFFORTLESS%20EUROPEAN%20REGISTRY
FACILITIES
PAYROLL
SALES%20SOURCE
WANT%20ADS

Lastly, when we reviewed the analytics we found some of the types of documents most frequently searched for by employees included policy documents and forms. So we proposed the creation of a dedicated section on the site where users could find and favorite (bookmark) frequently accessed items. Today a "Form & Policy Central" site enables users to quickly locate what they need faster than ever before. Adapting to the user's needs based on site usage data helped us implement a vastly improved site experience for our client.

Understanding International Use —
The Need for Multilingual Support on an Intranet

Figure 3 depicts how an organization's site visitors were spread across eight different countries. However, the site visit distribution was not representative of the actual employee distribution for these countries. International employees were not visiting the corporate intranet as frequently as U.S.-based employees. To overcome this challenge for the client, we had to determine the cause.

Figure 3. Global Intranet Site Use Distribution

Countries

8 COUNTRIES

USA	74%	AUS	2%
IRL	19%	CHN	1%
JPN	3%	Other	1%

Further analysis of the segmented data, along with user interviews, revealed that the majority of international users felt the information presented on the corporate site was not relevant to them. For example, they felt corporate and industry news and events were more targeted towards U.S. employees than to local international offices. Additionally, many users preferred content to be translated into their native language, even if it was about corporate events.

These insights were instrumental in the decision to create country-specific sites where relevant local content would have prominence. It was decided to provide users functionality to easily translate content into six different languages, such as Spanish, Chinese, Japanese, Portuguese, and French. These changes lead to a 35 percent increase in site usage for these different international locations. Several country, office, and business-unit specific sub-sites were also added to provide the most relevant and targeted information to the end user.

Social-Collaboration Features & Mobile Support—Upgrading to SharePoint 2013/Office 365

Using custom reports for specific features on several of our clients' intranets, we are easily able to measure the ROI for an upgrade or migration. One example of this is how we use information from analytics to prioritize current pain points and then map these to features available within SharePoint 2013.

This organization was in need of a better content management experience, social-collaboration capabilities, a federated search with non-SharePoint content, and an optimized mobile experience. These pain points were determined by using the data from Webtrends reports, and were mapped back to SharePoint 2013. Features such as flexible, meta-data-driven Information Architecture (IA), Communities/Groups, improved search capabilities, and integrated FAST and Device Channels for mobile and tablet devices were viewed as key wins and improvements for this organization.

Beyond just validating user needs and functionality on SharePoint 2013, we were also able to help this organization understand existing content usage to strategize their data migration efforts. A prime example of this was our ability to plan the archiving of content that was neither used nor updated in last 60 days, structuring data usage patterns into a logical architecture supporting maximum search and findability.

Improve It—Feature Enhancements and Proactive Governance

A measurement strategy doesn't stop when a solution is delivered. We understand that building a new system is only the first step in the process. As a result, we ensure our solutions are proactively supported and managed to continuously evolve and innovate. We do this in a number of ways. One of our key deliverables is a governance plan, which ensures our solutions not only remain stable and secure, but actually grow in usefulness over time.

Once launched, keep improving the system. Understand **usage** and **users**: What desirable actions do people take on the site? Can they complete their tasks faster or easier? **Develop goals** and success criteria, create a **measurement strategy** and **optimize** your intranet based on results.

The following are some examples of how to review the usability and success of our solutions and justify the need to enhance the overall user experience using analytics:

Improved UX and Design—UI Enhancements for Web Parts

One of our clients had initially implemented a central web part built using Adobe Flash. It was a multimedia carousel displaying the latest news and popular articles recently published by employees from across the globe.

The results from only a week's analytics on this feature lead to several discoveries, ultimately leading to its complete redesign. One of the most significant discoveries was that 12 percent of the site users either did not have the Adobe Flash browser plug-in installed or had it disabled on their devices. As a result, the carousel was redesigned using HTML5 and CSS3 to support a wider range of modern browsers and mobile devices.

This carousel also contained targeted video campaigns created by the internal communications department. Custom analytics reporting from Webtrends indicated how many times and how long video content was played. Upon data analysis, supplemented with additional user interviews, BrightStarr determined most users found videos were too small and ineffective for use.

As a result, the original design (Figure 4) was modified to support larger rich media and video content (Figure 5). Furthermore, news article content was moved from the media carousel into a separate web part, and placed into a tabbed section where the user could easily toggle between the *highest rated* and *most recently published* articles. The new design resulted in a 75 percent increase in the number of video views.

Figure 4. Video Wireframe Before Redesign Depicting Small Video Frame

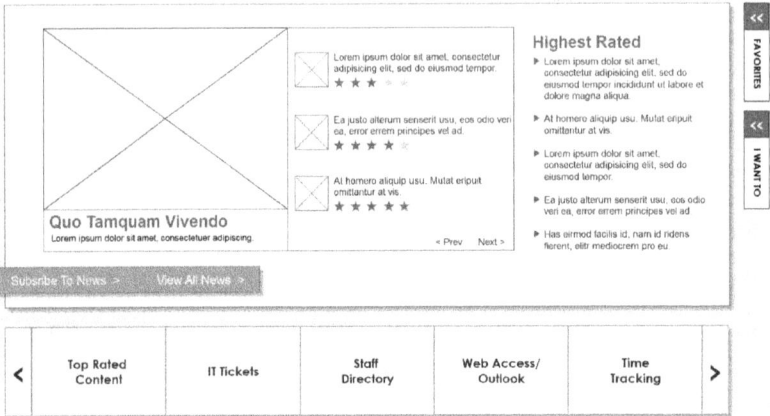

Figure 5. Redesigned Video Wireframe Depicting Large Video Frame

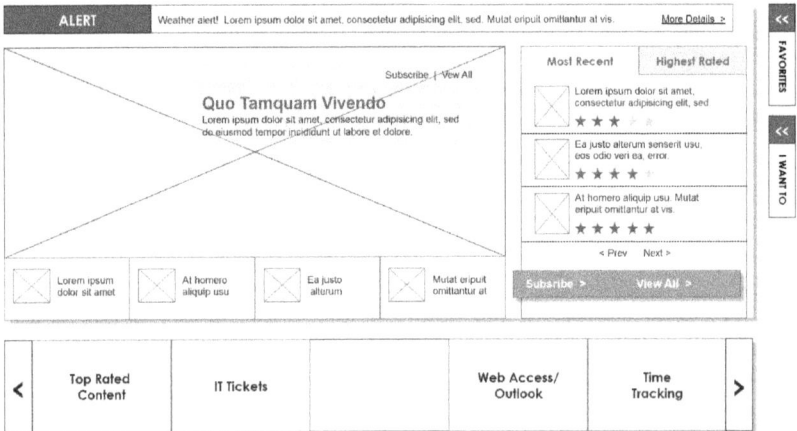

Another significant statistic pertaining to video content was that approximately 30 percent of intranet users stopped watching videos partway through. Analytics showed that the average user would either pause the video or exit the page within two minutes of starting any given video.

Originally the average video length was approximately four minutes, but when the video duration was reduced to two minutes the number of viewers who watched the complete video went up by 60 percent. This was a simple change that resulted in a significant increase in video views.

Ensuring Adoption and Streamlining Governance

A governance plan documents policies, roles, responsibilities and processes that control how an organization interacts with an IT system. Governance plans help streamline deployments, help protect against security threats or noncompliance issues, and ensure the best possible return on investment.

Webtrends not only helps us in planning new innovative features based on user behavior metrics, it also helps us develop adoption strategies once features have been rolled out. Training forms a significant part of our methodology for delivering successful intranets. For example, for certain frequently used web parts (Figure 6) we were able to create special training, as users were very keen on learning about all of their capabilities.

Figure 6. Frequently Used Web Parts

Web Parts by Page Views

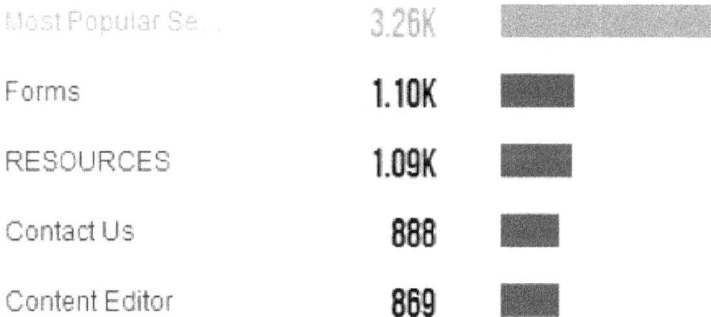

Most Popular Se...	3.26K	
Forms	1.10K	
RESOURCES	1.09K	
Contact Us	888	
Content Editor	869	

Special campaigns to introduce new features are also run throughout the site, with videos, articles and blogs describing the benefits of the features and how to use them. This strategy helps achieve a high adoption rate of new features, and raises the profile of the intranet as a whole. Webtrends usage statistics are instrumental in this process to approve any additions to the intranet, and carefully monitor existing features to ensure that they continue to add value.

Improving Navigation and Providing Engaging Content

Intranet users won't engage if the content isn't compelling or relevant. Intranet analytics can be used to identify the most and least popular content. So if you see that one of your posts isn't earning views, you could look to replace it with more interesting content, or improve its positioning. What if only 40 percent of employees viewed a required video message from your CEO posted on the intranet? Or if most employees visiting the intranet just view their paystubs? Webtrends SharePoint Analytics can help you clearly identify usage challenges, develop targeted strategies to improve end-user engagement, and benchmark performance to drive holistic organizational productivity.

Figure 7. Using Analytics to Provide Engaging Content

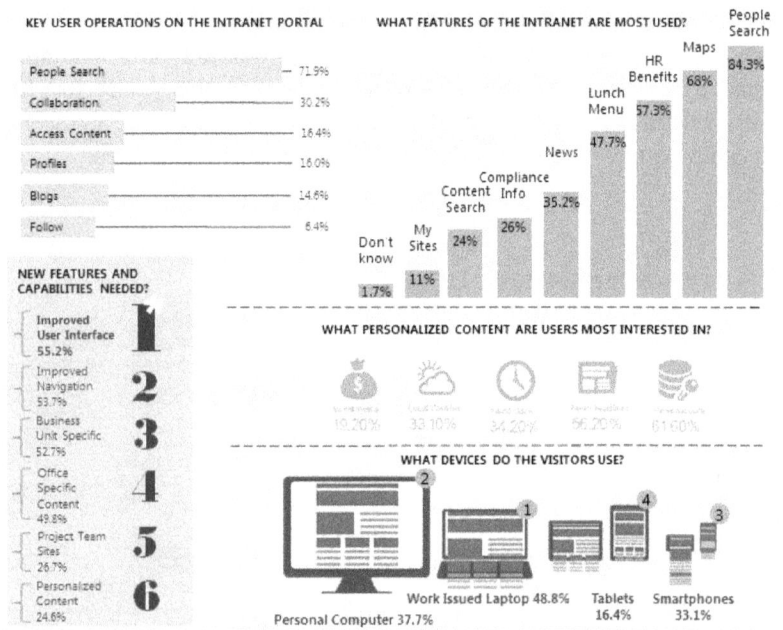

KEY USER OPERATIONS ON THE INTRANET PORTAL

People Search	71.9%
Collaboration	30.2%
Access Content	18.4%
Profiles	16.0%
Blogs	14.6%
Follow	6.4%

WHAT FEATURES OF THE INTRANET ARE MOST USED?

People Search 84.3%, Maps, HR Benefits 68%, Lunch Menu 57.3%, News 47.7%, Compliance Info 35.2%, Content Search 26%, My Sites 24%, Don't know 11%, 1.7%

NEW FEATURES AND CAPABILITIES NEEDED?

1. Improved User Interface 55.2%
2. Improved Navigation 53.7%
3. Business Unit Specific 52.7%
4. Office Specific Content 49.8%
5. Project Team Sites 26.7%
6. Personalized Content 24.6%

WHAT PERSONALIZED CONTENT ARE USERS MOST INTERESTED IN?

19.20% 33.10% 34.20% 56.20% 61.60%

WHAT DEVICES DO THE VISITORS USE?

Personal Computer 37.7% Work Issued Laptop 48.8% Tablets 16.4% Smartphones 33.1%

Evolving the intranet's navigation using analytics data is critical in ensuring adoption and sustaining it. It is very important for any user experience consultant to understand how users navigate to pages and content, how they interact, share and surface ideas, and which assets are no longer useful. SharePoint Analytics captures and reports on end-user activity, search effectiveness, content access and collaboration at the individual level. This insight enables you to understand how to best manage your communications protocols, assets and archival processes. Using the information from the analytics report, our consultants perform content audits and make navigation changes, ensuring users are able to complete their tasks with minimum effort.

Bridging Qualitative and Quantitative Data for Scale and Perspective

One of the strongest contributing forces behind being able to implement effective, cross-platform user experience strategies is partnership. Understanding the significant partnership between research data and design, as well as the partnership between the types of data analyzed and leveraged to properly inform design, is a key driver to UX success. The balance achieved when UX practitioners bridge and holistically measure both qualitative and quantitative data is unparalleled. This balance allows the entire experience with a system to be discovered, analyzed, placed in proper prospective, and continuously improved upon.

The distinction between what we call qualitative data and quantitative data can be summed up in two parts: what and why versus where and how. Quantitative studies focus more on indirectly determining where users are going and have gone, as well as how they are getting there. Whereas qualitative studies focus more on determining user behavior or attitudes by directly observing what users are doing and saying, as well as why they are performing certain actions.

Both sets of data can stand independent of one other and tell different sides of a story, but a proper experience can't be developed using only one side. In order to truly scale and put ones' entire user journey into perspective, both sets of data are needed and should be used together to validate one another.

Bridging qualitative and quantitative data may seem like common sense. However, depending on the project timeline, conflict often stems from both sets of data being on opposite sides of the scale, and being too difficult and/or time consuming to measure and analyze appropriately within one study.

Our methods—using surveying techniques, interviewing, observation and analytics—have proven how combining qualitative and quantitative data improves discovery and evaluation by ensuring the limitations of one type of data are balanced by the strengths of another. This idea further highlights not only the metaphor of two being better than one, but more importantly how effective data analysis is made up of more than mathematical data points and percentages. The key human element must be taken into account in order to provide an authentic and positive user experience.

There are a few options to choose from when bridging qualitative and quantitative methodologies. All these options, of course, are dependent upon project scope, timeline and resources. Some UX practitioners may have the bandwidth to collect both sets of data at the same time. Others may choose to gather each set one at a time in sequence.

The key to making the data most useful, regardless of the method used to gather, is ensuring that one set is used to help inform the other. This allows the research to be multivaried, and any design decisions made as a result to have multiple levels of supporting research. In addition, some practitioners may choose to combine their blended data constantly throughout a discovery/evaluation process, while others wait and combine their data at the end. The methodologies may vary, but the purposes remain unchanged.

In the simplest terms, bringing quantitative (the how) and qualitative (the why) data together helps to explain, examine or enrich one's overall user experience understanding. And in some cases (if you're lucky) it accomplishes all three.

Summary

BrightStarr has seen great success as a result of its partnership with Webtrends. Webtrends data provides a strong foundation on which to build innovative and effective solutions that directly impact our clients' needs. We have been able to use Webtrends to plan, build and execute effective solutions to help meet and exceed the client's success criteria in a quantifiable manner.

We have also had great success as a result of using a 360 approach to the data we analyze. By pairing qualitative and quantitative analytics, we are able to see the entire picture: current problems, current needs, current usage, etc. This partnership of data is incomparable in helping us both drive and teach effective user experience strategies. Being able to quantify and qualify one's experience with a system solution is a key driver, one we plan on continuing to use as our efforts in building award-winning, best-in-class intranets continue across the globe.

About BrightStarr

BrightStarr is a market leader in SharePoint and Office 365 consulting services. Specializing in the delivery of large-scale, end-to-end portal solutions, BrightStarr is proud to have already worked for some of the biggest names both in the private and public sector.

BrightStarr's innovative group of SharePoint experts have found tremendous success building award-winning intranets, extranets, and public-facing websites. BrightStarr has won the prestigious Nielsen Norman Intranet of the Year award thrice in the last four years. Maintaining Gold level competencies as a Microsoft Partner, BrightStarr has leveraged their continuing success to service a myriad of industries, from the retail sector to highly regulated areas such as healthcare and energy.

BrightStarr's uniqueness is the blend of both creative and technical skills that deliver truly extraordinary digital experiences with the perfect balance of form and factor. BrightStarr is an experienced and certified multi-year Webtrends implementation partner.

NOTES

[1] Gartner: http://www.gartner.com/newsroom/id/2867917

[2] Nielsen Norman Best Intranet Design Award–2012, 2014 and 2015: http://www.brightstarr.com/sharepoint-technology-and-application-insights/brightstarr-wins-nielsen-norman-best-intranet-design-award-2015

ABOUT THE AUTHORS

Kunaal Kapoor is Vice President of Business Development at BrightStarr. Kunaal is an experienced business and technology leader. He has years of professional experience in helping many of BrightStarr's strategic clients and ensuring smooth project executions.

Kunaal currently manages client partnerships and business development for BrightStarr. His expertise with technology and knowledge of user-experience best practices enables him to help business leaders leverage platforms like SharePoint, Office 365, and Yammer to enhance employee productivity, automate business processes and ensure ROI from their enterprise technology investments.

Kunaal has a unique ability to understand both business (end users) and technology to build best-in-class solutions. He also speaks frequently in SharePoint community events.

Antoinette Houston is Senior User Experience Consultant at BrightStarr. Antoinette is a highly motivated and detail-oriented UX professional. She has a hybrid and multi-faceted background in research, analysis, design (including strategy) and evaluation of technology solutions across all platforms (mobile, tablet and desktop).

As a senior UX consultant for Brightstarr, Antoinette plays a key role in driving and building positive user experiences, as well as effective and desirable designs. Antoinette's experience within both the private and public sectors has provided her with a unique range of skills. Her balance of positive relationship-building, effective communication styles and blended technical expertise, along with her consulting skills, help to ensure smooth project deliveries.

Antoinette has been a part of key projects for BrightStarr and has helped organizations understand their end-user needs to make strategic decisions on using platforms like SharePoint, Office 365 and Yammer to enhance adoption, productivity and ROI.

Chapter 12

Accomplish More with Webtrends Analytics for SharePoint

by Loren Johnson

As this book has shown, Microsoft SharePoint is a dynamic communications and collaboration solution offering a huge diversity of capabilities and user experiences. The best SharePoint deployments are vibrant and engaging, drawing in users from across the organization with intuitive connections, compelling experiences, and incomparable abilities to drive productivity. Often deployed as the central means for communications and collaboration in an organization, SharePoint is the conduit through which people and teams are bound together and can accomplish more.

SharePoint Evolution

Though its position in the organization may have evolved, SharePoint endures as an indispensable platform through which the exchange of ideas, information and innovation can be accomplished. Originally deployed to manage content and act as a hub to facilitate corporate communications, SharePoint has helped lead the way as the modern workplace has evolved. Microsoft has continuously updated SharePoint to stay ahead of the market, addressing such changes as:

- The need for accessibility from any device, anywhere, anytime for knowledge workers
- Increasingly collaborative and interactive work places
- Flatter corporate structures

As Jeff Shuey noted in his chapter, Microsoft has innovated SharePoint over its iterations to deliver an ever more dynamic, vibrant and productive marketplace of ideas. With aspirations to foster ever more organizational synergies, reveal connections, discover answers, and sync information, SharePoint is at the heart of many successful organizations.

The latest SharePoint iterations, SharePoint 2013 and Office 365, have baked-in social and collaborative capabilities, which like a market bazaar, enable the open exchange of creative thinking, discussions and ideas. Companies looking to take advantage of the best of their people and assets should be drawn to the broad communication capabilities in SharePoint 2013 and Office 365. Both versions allow companies to break down the barriers between people, offices, divisions and locations.

With SharePoint 2013 and Office 365, Microsoft has provided administrators and users ultimate control over the functional positioning of SharePoint within their organizations. Further innovations include the advent of the SharePoint app market and opening the SharePoint platform to third-party providers to offer purpose-built applications, easily integrated and designed to enhance functionality. Virtually unlimited customization through the app market enables SharePoint to deliver its unique capabilities to meet the business needs of its users more effectively than ever before, making SharePoint all the more vital to its buyers. This is not your grandfather's SharePoint.

Recently, Microsoft has begun divulging its plans for additional innovation in the SharePoint world. The company indicates the likely release of SharePoint Server 2016 in the next year or so, which will update SharePoint for those who require on-premises deployments. In this release, it is assumed that many of the latest capabilities found in SharePoint 2013 and Office

365 will be integrated for on-premises clients, including advanced social and collaborative resources, an integrated app market, and a more open platform. This will bridge older and newer SharePoint instances and allow for simpler management of hybrid environments.

Considering the thousands of SharePoint customers that have deployed multiple simultaneous versions and instances, Microsoft is wise in its plans to enable better hybrid deployment integration. Some companies have elected to invest in the newer capabilities of SharePoint Online to gain social and collaborative benefits. However, for many of them, moving off of a deeply integrated existing deployment can be problematic. These companies that straddle both worlds — with a SharePoint 2007 or 2010 on-premises deployment while using some elements of SharePoint 2013 or Office 365 online at the same time — may be positioned that way for many years. Even if a full migration to a newer version is planned, their hybrid environment status may be more permanent than temporary.

Whatever the deployment environment, it is clear that SharePoint customers look to Microsoft to continually improve SharePoint with increased flexibility, better connectivity, and broader integration. And as Microsoft continues to innovate SharePoint, many customers have been excited to invest in the new features and functions based on a promise of a more engaging, productive and powerful internal communications environment. After all, those new capabilities are designed to be used, to showcase the critical role internal communications can have, and to enable SharePoint advocates to prove its undeniable value to their organizations.

Prove It with Analytics

No matter what SharePoint version you use, understanding how it delivers value to the organization is critical. In our previous book, *Prove It! Using Analytics to Drive SharePoint Adoption and ROI*, we discussed the value of SharePoint measurement programs in delivering validation of the investment, site design and adoption strategies. As Kanwal Khipple showed in his chapter, these metrics are essential to setting a performance baseline against which key improvement tactics can be evaluated.

Forrester Research published its most recent analysis on the Share-Point marketplace, *Shift Your Expectations of SharePoint*, in December 2014. In this report, Forrester showed that the top two issues concerning

SharePoint not meeting buyer expectations remain users not liking the user experience and disappointing adoption levels. Other top concerns that respondents mention include not seeing business value from SharePoint and difficulty with migration strategies.

Best practices for driving end user engagement include:

- Employing complete user behavior metrics to baseline performance
- Setting objectives
- Ascertaining the impact of engagement tactics
- Making iterative improvements

Many companies have found that the more they invest in interactive and dynamic elements within their SharePoint, the more drawn to it users are—and they can prove these findings with data and analytics reporting.

These analytics reports can help define what tactics are most effective within teams or sites, especially when looking to improve engagement in a single SharePoint instance. Analytics form a foundation for validating conversations about what's next, what would best fit or improve an existing SharePoint environment, and how best to proceed. And as Jason Schnur discussed in his chapter, if you view SharePoint users as consumers with practical and emotional needs, you can ensure your engagement and measurement strategies are built around user communities realizing holistic value.

Improve SharePoint: User Experiences Define Success

Investing in a multi-instance SharePoint environment, a migration strategy, or an enterprise social network without a performance baseline virtually guarantees low buy-in, uncertainty and missed targets. As exciting as they are, any of these initiatives can effectively transform SharePoint into a dynamic marketplace of ideas and lively collaboration or, if executed poorly, an unruly cacophony of shouts and accusations. Ideally, well-defined targets and tactics would be part of any enhancement program, steering companies away from noise and dissonance and toward collective harmony.

As Susan Hanley wrote, to find critical moments of engagement in a social environment, it's important to think beyond simple collaboration to how ideation, sharing and productivity can blossom when a culture of social connectivity is cultivated into every facet of the workplace. Richard Harbridge mentioned that if you employ a methodical approach to managing end-user engagement in Office 365, you're more likely to see definitive results.

One central element threads through all of these use cases, chapters, and narratives. One component is vital to include in your plan for getting the most out of SharePoint and delivering the user experiences that drive ever more value for users, stakeholders and the organization as a whole. That crucial element is measurement.

The promise of SharePoint today is in its ability to deliver particularly engaging and dynamic sites, to draw users in and build preferences to use SharePoint first and finally. While these objectives may be met simply through implementation and internal viral marketing, the best deployments and the highest engagement rates result from implementing measurement strategies. Never stop improving.

What to Measure

Much of the demand for investing in SharePoint is focused on delivering users an environment where they can get more out of the company's assets and each other than ever before. Thus, user experience validation is a high-value metric. It's important to note however, that user experience means much more than the number of site visits, the total likes or follows, or how many documents a user downloads. Particularly in terms of business value and ROI, deeper and comparable engagement metrics are required.

User experience assessment normally results from a combination of multiple metrics, each of which tell part of the story. Yet, as Naomi Moneypenny highlighted in her social graphing chapter, once the organization understands the connection between user actions and other employees or documents, it can predict and anticipate follow-up activity, both positive and negative. And those insights can be exploited for further site enhancements. Patterns in aggregate user behavior can also emerge fairly quickly and can drive user experience enhancements.

Unlike driving adoption, which is a fairly linear process, improving user experience requires a continuous process of testing, analyzing, implementing and removing site elements. There is no singular point at which a goal is accomplished and the effort can stop. As the BrightStarr leaders wrote in their chapter, analyzing SharePoint user behavior is a continuous process, with users and advocates striving to continually improve the user experience. But the quality of the analytics engine makes a difference in enabling effectiveness. Depending on the ambition, the chosen analytics tool can grow with the organization as it drives to deliver continually better user experiences. Or it can limit progress and growth with an artificial limitation. These limitations in data accuracy, scalability, sophistication and reporting can impede an otherwise successful engagement and user experience improvement process.

Webtrends for SharePoint

As the central theme of this book is driving improvement in SharePoint through the use of measurement and analytics solutions, it's important to discuss which analytics features and functions deliver the most value. Some analytics tools capture data through the Microsoft API, but that data stream alone is insufficient to capture detailed user and action level metrics. Using JavaScript tagging on SharePoint sites can capture additional user behavior metrics and attributes. Combined, these data can deliver enough information to reveal basic SharePoint performance.

But to get the full picture of SharePoint performance, the analytics solution must:

- Be built specifically for SharePoint
- Capture all SharePoint-specific actions and user behaviors
- Not limit data capture or delivery
- Be able to present the data so key SharePoint stakeholders can easily understand and act on the reporting

There are several SharePoint analytics options on the market, including the Microsoft out-of-the-box SharePoint analytics solution. But only one solution combines multiple SharePoint-specific data sources, delivers the precision needed to drive engagement and user experience tactics anywhere within a SharePoint environment, and enables stakeholders to take action with confidence.

Webtrends Analytics for SharePoint is the market-leading measurement solution for SharePoint. Used by more than 300 global customers — including Microsoft — it is the only solution in the market built specifically for and Microsoft-preferred for every current version of SharePoint, including SharePoint 2010, 2013, and Office 365. Designed in cooperation with Microsoft and built in expectation of SharePoint product roadmap strategies, the Webtrends solution delivers what is needed for the SharePoint of today and tomorrow. Reflecting the ongoing commitment, Webtrends recently published an app for SharePoint 2013 and Office 365 in the Microsoft Store that makes analytics more accessible and more powerful than ever.

Microsoft is continuing to invest in SharePoint and in the solutions that are closely tied in with it in Office 365. New versions are on the way with more advanced communications and connectivity accessibility. Many companies are looking to invest in the latest capabilities, whether in hybrid or entirely new deployments, seeking to improve their internal communications potential.

Webtrends Analytics for SharePoint can help you:

- **Prove** the value of your investment,
- **Improve** performance continuously, and
- Make your SharePoint and your business **accomplish more, no matter what comes**.

ABOUT THE AUTHOR

Loren Johnson is a senior product marketing manager at Webtrends. He is responsible for all marketing associated with Webtrends Share-Point Analytics solution and contributes to its broader catalog of digital measurement solutions and capabilities. Prior to Webtrends, Mr. Johnson was a market analyst at Frost & Sullivan, where he produced analysis on digital analytics and content management systems. Loren writes thought leadership articles on the value of measuring productivity environments, including SharePoint. He can be reached at loren.johnson@webtrends.com.